Katharine V. (signature)

STUDENT'S GUIDE TO MASTERTON & SLOWINSKI'S

CHEMICAL PRINCIPLES

RAY BOYINGTON Department of Chemistry
University of Connecticut

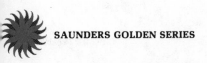
SAUNDERS GOLDEN SERIES

W.B.SAUNDERS COMPANY · PHILADELPHIA · LONDON · TORONTO

W. B. Saunders Company: West Washington Square
Philadelphia, Pa. 19105

12 Dyott Street
London, WC1A 1DB

833 Oxford Street
Toronto 18, Ontario

Student's Guide to Chemical Principles   ISBN 0-7216-1900-2

Print No: 9   8   7   6   5   4   3   2

PREFACE

This study guide has been written as an aid to the student of general chemistry — to supplement a textbook and lecture series or to guide independent study. The chapter sequence and selection of topics parallel that of *Chemical Principles*, 3d edition, by W. L. Masterton and E. J. Slowinski.

Given the general availability of problem-solving manuals (see the suggested titles at the end of this preface) and the abundance of worked examples found in most textbooks, the emphasis here is on the qualitative understanding and application of chemical concepts and theory, and not on problem solving, mathematical operations or memorization of formulas and equations.

Each chapter begins with a set of questions. These are meant to suggest some of the things you should be looking for in studying the topics of this chapter and the corresponding text and lecture material. Next follows a list of concepts and mathematical manipulations with which you should be familiar in order to get the most out of this unit of study. A chapter summary then presents an overview of some of the ideas treated in this unit. The major part of the chapter then follows in the form of a test, with answers immediately after. Finally, a brief annotated list of carefully selected and recommended readings is given — for more basic, background study; for interesting and important applications; and for in-depth study.

The major objectives, then, in writing this guide have been to provide you with tests by which you can measure your understanding of the material; to help you apply what you learn to new problems and to suggest further avenues of study; and to help you acquire a feel for the chemist's view of the physical world, an acquaintance with its limitations and its potentials.

SUGGESTIONS FOR THE USE OF THIS GUIDE

First, let's agree that what you get out of this course of study depends in great part on what you are willing to put into it. What follow are a few suggestions as to how you can make the most of your own efforts.

There are certain basic skills that you will frequently need to rely on, possibly further develop. The most fundamental of these are reading with comprehension and solving mathematical problems once they have been set up. Let's not worry about the first very much at this point. For one thing, it means learning new words and new meanings for old words. There is probably no better way to learn or improve a language skill than through repeated usage, and that is one reason for this guide. Also, the text problems and the tests in this guide will help you to decide just what you do and do not understand. As for the math, there are likely to be very few calculations in this course which you cannot do. These may consist of working with exponential notation; working with logarithms; solving first order (linear) equations; or solving less frequently encountered second order (quadratic) equations. For these you may simply need some review. Suggestions for background readings follow; additional help is generally yours for the asking, from fellow students and from your instructor. But here, it is a qualitative understanding, an "appreciation" of the ideas and the physical reality behind the math that we are after.

1. Attend lectures and take notes, at least at first. (You may eventually feel that lectures are a waste of time — you will need to find out for yourself.) Notes are important in answering questions such as these: What seem to be the important points? What is the lecturer's emphasis? What new examples and applications are given? What kind of experiments are done or cited?

2. Prepare for the lecture (assuming you go)! Read over the material you expect to be discussed. Try to get a general idea of what the material is about. (Some lecturers assume a basic preparation, some do not. Even if the lecturer tells you nothing new, you will have learned more by having done the work yourself.) The more you know about what the lecturer is saying, the more you will be able to learn; the fewer notes you will need to take, and the better they will be; the more questions you can raise; or, the more time you can devote to other studies.

3. Get as much as you can out of discussions with other students and with your instructor, out of discussion or problem-solving classes — by being prepared, and by understanding (first by discovering for yourself) the inherent limitations and potentials of your class. (You will generally find, for example, that an instructor is better able to explain specific points and answer specific questions than he is to guess what problems you are having.)

Make your problems known, as early and as clearly as possible. Take advantage of the skills and insights of your instructor and your fellow students! You usually do have a very real influence on how beneficial a discussion class may be. Remember that this guide is designed to help you evaluate your progress for yourself, to recognize those areas where you need help, and to assist you in finding it; but, it is no substitute for the classroom.

4. Review regularly and frequently, if only to avoid cramming for an exam. This is where your notes should be most helpful, and hopefully this guide as well. Keep this use in mind when taking notes from lecture or text. Look for connections between topics.

5. Read more chemistry — whether out of curiosity or from a feeling of helplessness. The more you do, the more rewarding this study must be.

To sum up:

Before attending lecture: Skim through the textbook chapter, reading introductory paragraphs and section headings, looking at tables and graphs; read and *add your own notes* to the first three sections of the study guide chapter. Then go back and *work* through the textbook chapter. In this and all other parts of the course, you've got to *actively participate.* Try to anticipate answers; write in questions you need answered; jot down your own conclusions in your own words.

Before discussing problems: Work through as many problems as possible; work the self-test in the guide; note anything requiring further explanation; think about your answers — do they seem reasonable in terms of what you already know? Do they suggest other problems?

Before taking exams: Review the main concepts and their physical meaning by referring to your notes and to the chapter summaries; recall and try to anticipate the emphases. Work some more problems.

For further assistance or enjoyment: Ask questions of your fellow students and your instructor; do some reading. Teach what you have learned.

ADDITIONAL STUDY AIDS AND SUGGESTED READINGS*

Math Preparation and Problem-Solving Manuals

Butler, I. S., and A. E. Grosser, *Relevant Problems for Chemical Principles,* Menlo Park, Calif., W. A. Benjamin, 1970.
  Lots of interesting problems with detailed solutions and commentary; often applied to realistic situations.
Masterton, W. L., and E. J. Slowinski, *Mathematical Preparation for General Chemistry,* Philadelphia, W. B. Saunders, 1970.
  Chemical problems are given throughout, with solutions; extensive discussion of math; of continuing usefulness.

---

*Besides magazine articles, readings suggested are generally available in paperback.

Peters, E. I., *Problem Solving for Chemistry*, Philadelphia, W. B. Saunders, 1971.
   Worked problems are "programmed" to encourage your participation; answers given for somewhat basic problems.
Pierce, C., and R. N. Smith, *General Chemistry Workbook*, San Francisco, W. H. Freeman, 1971.
   Solutions given to about half of the problems; includes discussion of some math.
Risen, W. M., Jr., and G. P. Flynn, *Problems for General and Environmental Chemistry*, New York, Appleton-Century-Crofts, 1972.
   Lots of problems of varying difficulty; solutions given to all.
Sienko, M. J., *Chemistry Problems*, 2nd ed., Menlo Park, Calif., W. A. Benjamin, 1972.
   Lots of worked examples; problems with answers; particularly well-written, straightforward; some math discussed.

Programmed Instruction

Barrow, G. M., and others, *Understanding Chemistry*, New York, W. A. Benjamin, 1969.
   Self-teaching, self-testing; good background and practice material for most of the topics covered here.

Journals and Magazines

Chemical and Engineering News
   C & EN is a weekly newsmagazine of the profession, published by the American Chemical Society; much is generally readable, from letters and editorials to special reports.
Chemistry
   A monthly variety of readings of diverse levels and quality; might be worth a subscription. (Chemistry, 1155 16th St., N. W., Washington, DC 20036; $6/yr or group rate.)
Journal of Chemical Education
   J Chem Ed is another professional journal; monthly; of some interest to beginning students, particularly for resource papers and latest approaches to teaching.
Scientific American
   Occasional articles of chemical or applied interest in this monthly magazine; generally of top quality.

Audio-Visual Aids

Tapes, films, slides, programmed instruction materials and other resources that might be used for self-instruction are often available but not used. Find out where they are and how to use them, and whether they are worth your time. Often available at any library. Reprints of published papers are often available free from the author; sometimes, for a small fee, from the publisher.

Handbooks

Two very useful reference books, primarily compilations of chemical and physical properties and frequently updated, are:
Handbook of Chemistry, *N. Lange, Ed., New York, McGraw-Hill.*
Handbook of Chemistry and Physics, *R. C. Weast, Ed., Cleveland, Chemical Rubber.*

ACKNOWLEDGMENTS

I would like to acknowledge some of the generous assistance given me in my own study of chemistry and in the preparation of this small book: Professor William L. Masterton, for his quick insight and straightforward teaching over many years, and for much writing and rewriting of this guide; Tom Chapin, for a student's view of the suggested readings; Barbara Braun, for typing much of the manuscript; and Barbara, for endurance.

* * *

This book is dedicated to you, the student of general chemistry. I welcome your criticisms and your suggestions for the many ways there must be for making this more useful.

CONTENTS

1 BASIC CONCEPTS

QUESTIONS TO GUIDE YOUR STUDY

1. What role does chemistry play in today's world? How might the role change in the near future? (Can you recall any issues recently debated in the news media?)

2. What kinds of problems do chemists try to solve? Are there any specially useful or simplifying approaches to their solution? Is there a chemists' "point of view"?

3. What kinds of materials does the chemist generally work with in the laboratory? (Are they, for example, the materials of the "real" world, like steel, plastic, dacron, beer, tobacco smoke?)

4. How does the chemist obtain, prepare, and purify the materials he studies?

5. What are some of the instruments and techniques of the chemist?

6. What properties or changes in properties does the chemist measure and use to describe materials? How are these measurements communicated to other scientists and to the public?

7. What limitations and uncertainties are there in these measurements? How are these communicated?

8. What kind of test would you perform in your kitchen to show that a sample of baking soda is "pure"?

9. What are some of the unsolved problems — the frontiers — in chemistry?

10. What questions would you want to ask of a government-employed chemist?

11.

12.

YOU WILL NEED TO KNOW*

Concepts

   Though no previous encounter with chemistry is assumed at this point, at least a general understanding is assumed for the notions of matter, energy, composition, experimental control, measurement . . . .

Math

   1. How to use exponential ("scientific") notation — Appendix 4.
   2. How to recognize and solve first order (linear) equations, like $y = mx + b$. — Readings.

CHAPTER SUMMARY—OBJECTIVES

   You know, perhaps all too well, that we live in an age of science and technology. An age that is now two to three centuries old, yet very much with us. For example, nearly ninety percent of all scientists that there have ever been are alive now. At least through the first half of this century science seemed, by any measure, to be growing exponentially: about every ten to fifteen years the number of scientific journals approximately doubled and the number of compounds known to chemists likewise doubled. (Try estimating the number of papers in chemistry expected in the year 2000, knowing that there were about 50,000 in 1950.)
   This course will introduce you to the underlying principles and methods of one of the more fruitful areas of scientific endeavor, an endeavor involving many thousands of persons in many nations. But just what does a chemist do?
   Most of us would probably agree that a chemist is one who is qualified to determine the feasibility of physical (that is, not spiritual) change, of chemical reaction. A major objective of the chemist is to be able to predict the conditions under which a chemical reaction may occur and to describe the course the reacting system may take. (Does this suggest the kind of social role the chemist might play? After more than ten years of research, and applying a well known chemical principle, F. Haber was able to describe the

---

   *All chapters and appendices referred to are in *Chemical Principles;* otherwise, see the readings list for appropriate background material.

conditions under which atmospheric nitrogen could be converted to ammonia. A process made commercial in 1913, the Haber preparation accounts for most of the ammonia produced, whether for eventual use in fertilizers, explosives, or Windex. with Ammonia-D.)

A major objective for the student in an introductory course is to see how the principles of chemistry are experimentally established and then applied in predictions of reaction feasibility. The chemical systems that we will look at will be simple in composition (that is, there will generally be only one or two components) and simple in the number of things that happen (only one physical or chemical change will be considered). The chemist is not alone in this process of analyzing, of looking at one thing at a time; on the other hand, the chemist does always come round to standing back and taking an overview.

The history of modern chemistry really began with the introduction of quantitative experimentation, with measurement. The idea of composition, barely mentioned here, is taken up in more detail in the next two chapters. There are several levels of meaning to *composition*: one speaks of atomic composition, for example; or elemental composition, or composition by weight, and so forth. All of these usages refer to the way in which component parts or building blocks, whether simple in themselves or very complex, are put together to form matter as we know it.

A first chapter or lecture in chemistry is often devoted to a brief examination of the tools and units of measurement commonly used by the practicing chemist. Hence, once more the student hears all about the metric system. But this time for good reason: the chemist consistently uses the metric system and nothing else.

Objectives

Before leaving the material of this chapter, you should make sure that you have a working knowledge of the following:

1. be able to give operational definitions for substance, element, compound (that is, be able to describe actual processes that would isolate and identify);

2. be able to describe a variety of methods for isolating and identifying pure substances;

3. be able to work with conversion factors, particularly within the metric system (Note that *all* the problems worked in the text employ the unit conversion factor approach, or dimensional analysis);

4. be able to work with exponential notation and with significant figures.

SELF-TEST

True or False

1. Changes in physical state, like melting and boiling, tend to resolve matter into pure component substances. (T)

2. Extensive properties characterize, chemically identify, a substance. (F)

3. In trying to identify a certain liquid compound "L", a student finds that its density, freezing and boiling points, absorption spectrum, and behavior in a chromatography column are indistinguishable from those of known compound "Z". The student can safely assume L and Z are one and the same compound. ( )

4. Nearly all countries, including England and excepting America, have already adopted or are now adopting a metric system as the single recognized system of measurement. (T)

5. The simplification in using the metric system is that, when converting units within the system, a decimal point is moved. (T)

6. Appropriate conversion factors would allow you to convert from a volume measurement in cubic feet to a density measurement in grams per cubic centimeter. (T)

7. Since silicon is the second most abundant element in the earth's crust, it must be one of the cheapest to buy from a chemicals supplier. (F)

8. Sodium chloride and potassium dichromate, both solids which are soluble in water, can probably be separated by taking advantage of differences in solubility. (T)

9. The symbol for an element is always derived from the first one or two letters in its English name. (F)

Multiple Choice

10. One way to definitely show that gasoline is a mixture of substances would be to (C)
    (a) measure its density
    (b) measure the temperature during boiling
    (c) burn it
    (d) filter it

11. A procedure appropriate to the separation of the components ($\mathcal{C}$) of gasoline is:
    (a) burning   (b) fractional crystallization
    (c) fractional distillation   (d) column chromatography

12. The fact that behavior on melting may allow you to decide ( ) whether or not a sample of matter is a pure substance depends on the general observation that
    (a) melting points can be measured with high accuracy
    (b) all pure substances can be melted
    (c) the melting temperature is sensitive to small amounts of impurities
    (d) the melting point does not vary with sample size

13. A sample of matter that exhibits uniform behavior during all ( ) changes in physical state would have to be
    (a) an element   (b) a compound
    (c) a mixture   (d) a pure substance

14. Which one is an example of an extensive property? (A)
    (a) volume   (b) color   (c) temperature   (d) density

15. A suitable, nondestructive test for determining whether or ($\mathcal{C}$) not a particular beautiful, green gem is an emerald would be:
    (a) a melting point determination   (b) see if it is scratched
    (c) measure its absorption spectrum   (d) weigh it

16. Identification of the protein components in Super Crunchy ( ) Corn Flakes would probably involve
    (a) distillation   (b) recrystallization
    (c) filtration   (d) chromatography

17. For a given substance, density ordinarily increases in the (C) order:
    (a) solid, liquid, gas   (b) liquid, solid, gas
    (c) gas, liquid, solid   (d) solid, gas, liquid

18. If radioactive dating reveals the age of a manuscript to be ( ) between 1150 and 1390 years, its age may be represented as
    (a) 1.270×10^3 yr   (b) 1.27×10^3 yr
    (c) 1.3×10^3 yr   (d) 1×10^3 yr

19. How many significant figures are there in the number $6.50 \times$ ( ) 10^3 ?
    (a) two   (b) three   (c) four   (d) six

20. If the volume and mass measurements on a sample of arsenic   ( )
are 2.10 ml and 12.040 g, the reported value for the density of
arsenic, in g/ml, should have how many significant figures?

    (a) two   (b) three   (c) four   (d) five or more

21. How many times larger is a Kelvin degree than a Fahrenheit   ( )
degree?

    (a) 1.8   (b) 1/1.8   (c) 32   (d) they are the same

22. It is found that the chirping frequency of a tree cricket, f, is   ( )
approximately related to the Celsius temperature by the equation:
$°C = 0.1\,f + 8$. One would expect that the chirping frequency and the
Fahrenheit temperature would be

    (a) unrelated
    (b) equal to each other
    (c) related by the equation for a straight line
    (d) directly proportional to each other

SELF-TEST ANSWERS

1. **T**   (The basis for fractional distillation, for example.)
2. **F**   (Intensive properties do.)
3. **T**   (Identity of properties must mean identity of composition.)
4. **T**
5. **T**
6. **F**   (Volume and density are different kinds of measurement.)
7. **F**   (It is not very readily isolated from its compounds.)
8. **T**   (The basis for fractional crystallization.)
9. **F**
10. **b**   (A constant temperature would be characteristic of a pure substance, as a general rule.)
11. **c**   (The bulk of the mixture consists of liquid substances.)
12. **c**
13. **d**   (Either a compound or an element would behave this way.)
14. **a**
15. **c**
16. **d**   (See Chapter 23.)
17. **c**   (Water is exceptional; Chapter 11.)
18. **c**   (The second digit is uncertain by one unit; the third, even more.)
19. **b**
20. **b**
21. **a**
22. **c**   (Try substituting $0.1\,f + 8$ for $°C$ in: $°F = 1.8\,°C + 32$.)

SELECTED READINGS

Bachmann, H. G., The Origin of Ores, *Scientific American* (June 1960), pp. 146-156.
 A discussion of the abundance and availability of elements; an introduction to geochemistry.
Hammond, G. S., Relevance of Pure Research Today, *J. Chem. Ed.* (June 1971), pp. 362-364.
 One scientist's view of some current problems within the sciences and between science and society.
Keller, R. A., Gas Chromatography, *Scientific American* (Oct. 1961), pp. 58-67.
 A closer look at a very important analytical and preparative tool.
Kistiakowsky, G. B., American Science at the Crossroads, *C & EN* (April 24, 1972), pp. 30-33.
 Another short but provocative introduction to some current problems, particularly about the social responsibility of scientists.
Masterton, W. L. and E. J. Slowinski, *Mathematical Preparation for General Chemistry*, Philadelphia, W. B. Saunders, 1970.
 A good place to go for math background, practice exercises.
Price, D. J. de S., *Little Science, Big Science*, New York, Columbia University Press, 1963.
 On the exponential growth of science and its possible implications.
Ritchie-Calder, Lord, Conversion to the Metric System, *Scientific American* (July 1970), pp. 3-11.
 Light reading on the history and adoption of the metric system.
Ziman, J. M., *Public Knowledge*, Cambridge, Cambridge U. Press, 1968.
 Discusses the interactions between the scientist and his colleagues; or, how the scientific community arrives at a consensus and calls it scientific knowledge. Readable though not easy.

2 ATOMS, MOLECULES AND IONS

QUESTIONS TO GUIDE YOUR STUDY

1. Can you think of common, everyday observations which suggest that matter is made of atoms and molecules; or that it isn't?

2. If matter were not composed of atoms and molecules, would all chemists be unemployed?

3. Why do neutral atoms exist? (Why not a world made up of ions?) Or, why do ions form, some bearing positive charge, others negative?

4. How would you experimentally show that sulfur and oxygen combine in a one-to-one weight ratio to form the gaseous compound sulfur dioxide?

5. Just how small are atoms and molecules? Is there a convenient way of counting them; of expressing their numbers?

6. How is the mass of an atom or molecule determined? If they are too small to work with individually, then how do we know their masses?

7. If more than one kind of atom exists for a particular element (e.g., strontium-90 is but one kind of strontium atom), then how are we to interpret the numbers representing the weights of atoms?

8. What holds atoms together in a molecule? (Could you explain it to a first grader?) (Chapter 7 gives a detailed discussion.)

9. What are the supporting arguments, based on observation, for the atomic theory of matter?

10. If the idea of atoms is at least as old as ancient Greece, then why did it take so long for the atomic theorists to get anywhere?

11.

12.

YOU WILL NEED TO KNOW

See this section of Chapter 1, Study Guide.

CHAPTER SUMMARY—OBJECTIVES

With more than a little help from his friend the physicist, the chemist has solved the problem of the nature of bulk matter and the changes it undergoes. He interprets ("explains") the unique properties of substances, their descriptive and characteristic qualities, in terms of the unique properties of invisibly small particles that consitute matter: atoms, ions and molecules. The directly observable behavior of matter is the collective behavior of very large numbers of these particles. This imaginative view of matter provides the framework for most of the explanation, and success, of modern science. The biologist extends this theory to deal with very large molecules; the geologist, to account for the large-scale processes occurring within the earth; the astronomer, to explain the birth and death of stars.

It is a long way to go from an individual molecule of water to a clean mountain brook, or even a beaker of distilled water; but all the fundamental principles appear to be known. These principles comprise "atomic theory" or "kinetic molecular theory," and were established between 1800 and 1930, approximately. During that time, and since then, the theory has remained dynamic — growing in its detail and sharpness of focus with every new experiment. Atomic theory is the unifying theme of this course.

Now, the chemist certainly doesn't think of atoms as "ultimate" particles; but atoms, and to a certain extent, part of their underlying structure, are as fundamental, as far down the ladder, as he need consider. (The high energy reactions of modern physics, well outside the range of reactions considered to be chemical, have turned up two hundred subatomic particles, or thereabouts. Except for electrons and nuclei, our only encounter with such interesting and extraordinary matter will be to mention neutrons — and then only to explain the fact that atomic weights must be considered as average values, and to find that low energy, "chemical" neutrons sometimes bring about interesting reactions. More about this latter topic in Chapter 22.)

The arrangements of electrons and nuclei in atoms, ions, and molecules form the theoretical basis of all chemistry. Their properties seem sufficient for explaining bulk samples of matter. All observable changes are thought of as being changes in the arrangements of electrons and nuclei within atoms, ions and molecules; or as rearrangements of the atoms, ions and molecules themselves; or both. (This elaboration of atomic theory begins mainly with Chapter 6 and continues throughout the text.)

The laws of chemical combination (all of the important ones, like the law of constant composition, were known over a century ago provide some rather indirect support for these ideas. More direct and more compelling support for atomic theory, including the existence of particles smaller than atoms, comes from relatively recent experiments: electrical discharge through gases; atomic and molecular spectra (Chapter 6); x-ray diffraction by crystalline solids (Chapter 9); and radioactivity (Chapter 22).

How do you measure out a number of material objects, say three dozen apples? Everyday measurements would include those of simply counting, weighing, and determining volume. A small number of apples you no doubt would count off. But what if, for your pie factory, you needed three thousand? You would probably weigh them out. Conceivably, you could even measure their volume – a bushel of apples anyone? Well, these same three kinds of measurement are the only ways in which the chemist can normally deal with bulk matter consisting of vast numbers of particles. One of the most important things for him to know, then, is the relationship between mass and volume of a sample and the number of particles the sample contains. Volume measurements are typical in the case of handling gases (Chapter 5) or liquids and solutions (Chapter 10); weighing is the only other everyday measure. Even the largest molecules are too small to be counted: there is no chemical counterpart to everyday counting. Hence the need for relating masses and numbers of particles.

The gram atomic weight of an element, in particular of isotope ^{12}C, is a convenient reference sample for measuring mass and determining the number of particles present in the sample, all at the same time. Several independent methods of determining the number of atoms in exactly 12 grams of ^{12}C give a value of 6.02×10^{23} (see Readings for determining N). Avogadro's number is to the chemist what the ream of paper is to the typist, and the dozen to the poultry farmer. It is the number of atoms found in one GAW of *any* element; the number of molecules in one gram molecular weight of *any* substance composed of molecules. It is the number of structural or formula units known to have a certain mass, a gram formula weight.

Another need or use for this relationship is to be found in the way the chemist finds it convenient to represent reactions: the chemical equation, as you will see in the next chapter, treats the reaction of bulk matter as though individual atoms, molecules, and ions were taking part, step by step, to give the overall, net or observable, result. Equations show how we imagine counting off particles, and this we have done by measuring a mass.

Objectives

Before going on to the next chapter, you should be able to:
1. define element and compound according to the atomic theory;
2. distinguish between atom, ion and molecule;
3. give the approximate dimensions (say, diameter and mass) of atoms and their component parts;
4. relate gram atomic and gram molecular weights to masses and number of particles (e.g., calculate the mass of any given number of molecules, given the numbers and kinds of atoms present in one molecule).

SELF-TEST

True or False

1. The chemical properties of an atom are determined by its nuclear charge. ( )

2. In the light of current knowledge, we must view the law of constant composition as only a good approximation to observed behavior. ( )

3. An isotope is one of two or more atomic species having the same atomic number but different numbers of electrons. ( )

4. The following statement is probably consistent with modern atomic theory: a positively charged sodium ion is smaller than a neutral sodium atom. ( )

5. The molecular weight of a substance is simply a number which tells how heavy a molecule is when compared to a chosen reference. ( )

6. A gram molecular weight of hydrogen (two atoms in a molecule) is a gram of hydrogen. ( )

7. The number of molecules in a gram molecular weight of water, 18 grams, is eighteen times the number of atoms in a gram atomic weight of hydrogen, 1 gram. ( )

8. All neutral atoms of a given element have the same number of electrons. ( )

9. Stable, bulk samples of matter carry little or no electric charge; that is, they are electrically neutral. ( )

10. The atomic weight is an average number that takes into ( )
account all known isotopes of an element, including those prepared
artificially in the laboratory.

Multiple Choice

11. The atomic weight of chlorine is 35.5. This means that ( )
   (a) the actual weight of a C1 atom is not known more
       accurately than to three digits
   (b) a variable number of electrons gives a fractional value
   (c) on the average, an atom of C1 weighs almost three times
       as much as an atom of carbon
   (d) chlorine occurs with a variable number of protons

12. Suppose the atomic weight scale had been set up with ( )
calcium, Ca, chosen for a mass of exactly 10 units, rather than about
40 on the present scale. On such a scale, the atomic weight of oxygen
would be about
   (a) 64   (b) 32   (c) 16   (d) 4

13. The most direct method for determining atomic weights is: ( )
   (a) gas density measurements   (b) mass spectroscopy
   (c) combining weights   (d) α-particle scattering

14. An element is: ( )
   (a) a collection of atoms with identical numbers of neutrons
   (b) a collection of atoms with identical nuclear masses
   (c) a collection of atoms with identical nuclear charges
   (d) one of the following: air, earth, fire, water

15. Experimental support for the existence of atoms and sub- ( )
atomic particles includes all of the following except:
   (a) radioactive decay
   (b) electrical discharge through gases
   (c) metal foil scattering of alpha particles
   (d) continuous mechanical subdivision of a single crystal

16. By about how many orders of magnitude (powers of ten) is ( )
the diameter of an atom larger than that of a nucleus?
   (a) 1   (b) 2   (c) 4   (d) 10,000

17. The law that relates the different weights of one element that ( )
combine with a fixed weight of another element is the law of
   (a) conservation of matter   (b) constant composition
   (c) multiple proportions   (d) equivalent proportions

18. There is always a ratio of small whole numbers between:　( )
   (a) the gram atomic and gram molecular weights of an element
   (b) the weight percentage of copper in any two of its compounds
   (c) the weights of copper combined with one gram of element A in CuA and one gram of element B in CuB.
   (d) the atomic weights of any two elements

19. The diameter of an atom of gold is of the order of　( )
   (a) 10^{-5} cm　(b) 10^{-8} cm　(c) 10^{-12} cm　(d) 10^{-23} cm

20. The mass of an individual atom is of the order of　( )
   (a) 10^{-22} kg　(b) 1/2000 g　(c) 10^{-22} g　(d) 6×10^{23}g

21. The idea that most of the mass of an atom is concentrated in　( )
a very small core, the nucleus, is a result of the experiments of
   (a) Dalton　(b) Bohr　(c) Cannizzaro　(d) Rutherford

22. The charge on the nucleus of a neon atom is:　( )
   (a) +20　　(b) +10　　(c) zero　　(d) −10

23. The reason that the densities of two different gases at the　( )
same temperature and pressure compare as their respective molecular weights is:
   (a) all gases have the same molecular weight
   (b) all gases have the same density at the same temperature and pressure
   (c) all gas molecules are monatomic
   (d) assuming Avogadro to be right, each gas density is directly proportional to the mass of the respective gas molecule

24. Sufficient information for your calculation of the simplest　( )
atom ratio of nitrogen to hydrogen in the compound hydrazine would be:
   (a) the atomic weights of nitrogen and hydrogen
   (b) the ratio of masses for nitrogen and hydrogen atoms and the composition by weight of hydrazine
   (c) the combining volumes of nitrogen and hydrogen in forming hydrazine
   (d) the atomic weights of nitrogen and hydrogen, and the weight composition of several other nitrogen-hydrogen compounds

25. In one gram molecular weight of hydrogen, H_2, there are ( )
Avogadro's number of:
- (a) hydrogen atoms
- (b) electrons
- (c) hydrogen molecules
- (d) neutrons

SELF-TEST ANSWERS

1. T
2. T (We do find some atom ratios are not ratios of small whole numbers and that some ratios are at least slightly variable.)
3. F (Different numbers of neutrons.)
4. T (As you might guess, removing one or more electrons subtracts from the total volume of the atom, which is mostly electrons anyway.)
5. T
6. F (2.0 grams.)
7. F (There are 6.02×10^{23} molecules in one GMW, 6.02×10^{23} atoms in one GAW.)
8. T
9. T (This observation is often called the principle of electro-neutrality.)
10. F (Only naturally occurring isotopes.)
11. c
12. d (The weights of the individual atoms still compare 40/16, or 10/4.)
13. b
14. c
15. d
16. c (Atomic, 10^{-8} cm; nuclear, 10^{-12} cm.)
17. c
18. a (Molecules contain whole numbers of atoms.)
19. b
20. c (Divide any GAW by N.)
21. d
22. b (Same as atomic number.)
23. d (For a given volume, you are comparing equal numbers of molecules.)
24. b
25. c

SELECTED READINGS

Causey, R. L., Avogadro's Hypothesis and the Duhemian Pitfall, *J. Chem. Ed.* (June 1971), pp. 365-367.
  How and why Avogadro's idea came not to be accepted for some fifty years, even by the best of his colleagues.
Feinberg, G., Ordinary Matter, *Scientific American* (May 1967), pp. 126-134.
  A survey of the history of ideas about the structure of matter.
Hawthorne, R. M., Jr., Avogadro's Number: Early Values by Loschmidt and Others, *J. Chem. Ed.* (Nov. 1970), pp. 751-755.
  A historical comparison of several methods of determining N.
Lagowski, J. J., *The Structure of Atoms,* Boston, Houghton Mifflin, 1964.
  Useful here and for Chapter 6; presents excerpts and comments on experimental support for atomic theory.
Lavoisier, A., *Elements of Chemistry,* New York, Dover, 1965.
  A principles approach to chemistry as known in 1789; used by Dalton in his teaching. It contains the first(?) statement of the law of conservation of matter.
Lucretius, *On the Nature of the Universe,* Baltimore, Penguin, 1951.
  A two thousand year old classic that addresses you in the framework of a highly speculative atomic theory.
Patterson, E. C., *John Dalton and the Atomic Theory,* Garden City, N. Y., Doubleday-Anchor, 1970.
  A biography that is interesting for its re-creation of the social and scientific setting for Dalton's contributions.

3 CHEMICAL FORMULAS AND EQUATIONS

QUESTIONS TO GUIDE YOUR STUDY

1. What information is conveyed by a chemical *formula;* a chemical *equation?* (What is the distinction between the chemist's and the layman's use of the word *formula?*)

2. Is there just one kind of chemical formula; chemical equation?

3. What kind of experiment would you do to show that the formula of water is H_2O, and not HO as Dalton believed? What assumptions, if any, do you need to make?

4. How are chemical formulas and equations dependent upon a knowledge of the masses of individual atoms and molecules?

5. How does the chemist conveniently deal with numbers of reacting molecules; with their masses?

6. How do you represent the physical state (solid, liquid, solution . . .) of the materials taking part in a reaction?

7. Are the conditions, like temperature and pressure, under which a reaction occurs represented by the chemical equation for the reaction?

8. What, if anything, does a chemical equation tell you about what you would see as a reaction proceeds? (For example: what shape, size and color crystal is formed in a certain reaction, and how fast?) Or about what the molecules are doing?

9. What happens when you use nonstoichiometric amounts (quantities other than those represented by the chemical equation) in carrying out a reaction?

10. Does the chemist establish the chemical equation for each reaction he discovers? What kinds of observations does he make?

11.

12.

YOU WILL NEED TO KNOW

Concepts

1. The meaning of gram atomic weight, gram molecular weight — Chapter 2

2. How to interpret formulas; in particular, the meaning and use of subscripts, parentheses and brackets. These ideas are implicit in Chapter 2 — see Chapter 3 Summary and Self-Test, Study Guide; Readings.

Example: The formula P_4O_{10} refers to a molecule in which there are *four atoms of phosphorus combined with every ten atoms of oxygen.* As you will see in this chapter, the formula itself may refer to just one unit of structure, a molecule in this case, or it may refer to Avogadro's number of structural units — depending on the context in which the formula is used. But in any context, the atom ratio is four to ten in the structural unit.

The formula $Ca(NO_3)_2$ refers to a unit of structure in which there are *one Ca atom, two N atoms, and six O atoms;* or to Avogadro's number of such formula units. (The parentheses set off atoms which together may act as a unit; the subscript 2 then refers to two such groups of atoms.)

Math

1. How to work with conversion factors — Chapter 1
2. How to deal with significant figures — Chapter 1
3. How to relate gram formula weights (GAW, GMW), masses and numbers of particles — Chapter 2
(Examples: What is the mass of 2.5 GMW of acetic acid, $H_4C_2O_2$?
How many molecules are there in this much acid?)

CHAPTER SUMMARY—OBJECTIVES

Stoichiometry is the awesome label usually attached to the arithmetic of chemistry. It is the quantitative interpretation and use of chemical formulas and equations. The practical importance of stoichiometry is easily illustrated — the principles established in this chapter will be applied throughout the text and in any quantitative laboratory work you do. And beyond this course: the mining engineer may want to know the amount of iron he can expect to extract from a given amount of the ore hematite, Fe_2O_3; the botanist may wish to determine the volume of oxygen liberated when a certain mass of glucose is photosynthesized; the weight-watcher may want to compare the energies stored in equal masses of carbohydrate and fat.

This chemical arithmetic is not particularly difficult or complicated but it is likely to be new to you in one or two ways. Its mastery will require considerable practice. First, it involves different labels and units: grams, formula weights, moles, and others. Second, the calculations will always be carried out using conversion factors instead of proportions (which many people would rather use, if only out of habit). This work-saving "mole method" of solving stoichiometric problems involves a consistent interpretation of all amounts of chemical species (atoms, ions, molecules, electrons, . . .) in terms of a single unit, the *mole.* (And, in contrast to the use of proportions, it is more widely applicable.)

A mole measures a unit amount of a single chemical species or substance. This means two things: a mole is a *number* of particles; a mole represents a definite *mass,* associated with this number of particles.

(1) A mole is a counting unit with exactly the same kind of meaning and uses as other, more familiar counting units like *dozen.* A mole is Avogadro's number of identical items; the dozen, twelve items. Where we have used the words "gram atomic weight," "gram molecular weight," and "gram formula weight" we now substitute the one word mole.

(2) A mole is thus a variable unit of mass, its value (in grams) depending on the formula weight.

Consider that a "dozen identical objects" could just as well mean, simultaneously, a certain mass of objects and their number. If a penny weighs 3.0 grams, then counting out a dozen pennies is equivalent to weighing out 36 grams of pennies. Any amount of pennies, specified in terms of the unit dozen, will automatically mean a certain number of pennies and a certain mass of pennies. A unit amount (i.e., a dozen) of dimes will contain the same number of coins as a dozen pennies but will have a different mass (about 27 grams). We could conveniently count *or* weigh any amount of identical coins in units of dozen. In particular, we could be sure that, say, 72 grams of pennies and 54 grams of dimes contained equal number of coins.

The mole concept provides the chemist with the means of obtaining equal numbers of formula units by making the convenient measurement of weighing. For two substances, he needs to weigh out masses that compare in the same way as do the formula weights. (Of course, the numbers of particles desired may not be equal: you would simply weigh out masses such that the *numbers* of gram formula weights, or moles, compare in the same way as do the desired numbers of particles or formula units. For example, satisfy yourself that 71 grams of chlorine, Cl_2, contain twice as many molecules as does 1.01 grams of hydrogen, H_2.) The real advantage of working with moles is that with large-scale samples of matter we are dealing with the same numerical ratios of particles that are thought to react at the level of individual atoms and molecules. (A chemical equation can be taken to mean, for example, the relative numbers of particles reacting *or* the relative numbers of moles of particles reacting.)

The mole method of solving problems might be outlined as follows:

(1) Write the (balanced) chemical equation for the reaction.

(2) Use the coefficients of the equation as conversion factors for relating the numbers of moles of given and desired species.

(3) Convert the given amounts to moles by using the appropriate gram formula weights (i.e., grams per mole).

(4) Convert from moles of desired species to mass, if needed, by using gram formula weights, or to numbers of particles by using Avogadro's number (i.e., particles per mole), thus expressing the result in the units called for.

More often than not, the units of the desired quantity will indicate the kind of calculation you need perform. Suppose, for example, that you are given the mass of a reactant and are asked to find the mass of a certain product in a reaction. The series of conversions must take you from "grams reactant" through a relation of the numbers of moles of reactant and product, finally to the "grams product." The conversions called for in applying the mole method would look like this:

$$\text{g product} = \text{g reactant} \times \underbrace{\frac{1 \text{ mole reactant}}{\text{GFW reactant}}}_{\text{I}} \times \underbrace{\frac{\text{no. moles product}}{\text{no. moles reactant}}}_{\text{II}} \times \underbrace{\frac{\text{GFW product}}{1 \text{ mole product}}}_{\text{III}}.$$

Note that step I converts grams to moles; step II is the ratio of the coefficients gotten from the balanced equation; step III gives the desired units, converting from moles to grams.

Objectives

After studying this chapter of the text, you should be able to:

1. interpret formulas and equations in terms of relative numbers of particles, in particular, in terms of moles;

2. quantitatively relate moles, masses and numbers of formula units or particles (e.g., calculate the number of moles, given the number of grams);

3. calculate the simplest formula form composition data, and the composition, given the formula;

4. begin writing balanced equations for reactions described in the text and encountered in your experimental work.

NOTES ON WRITING CHEMICAL EQUATIONS

It may appear that you are being asked to write and understand chemical equations for lots of reactions when you don't yet know any chemistry. How, for example, do you know which substances are gases, which are solids? Or which substance is composed of molecules and which is not — and so what kind of formula to use? Or what conditions are required for the reaction to proceed as indicated by the equation? Our main concern at this point is with establishing a systematic approach to solving stoichiometric problems, and this does require at least some writing of balanced chemical equations. *The descriptive chemistry will be learned as you go along.*

So, what information do you need to be given and what do you need to find on your own — and where — so as to write equations? The following suggestions are to assist you in answering these and other questions. Further practice and suggestions can be found in looking ahead to Chapters 16, 18, 19 and 20, as well as to the Readings.

An equation represents, both qualitatively and quantitatively, observed changes in matter in terms of the rearrangements of atoms, ions, and molecules. An equation is "balanced" when it reflects the observed mass relationships between reactants and products, in particular the conservation of matter. Balancing requires that the same kinds and numbers of atoms appear in the products as started out in the reactants. To write a chemical equation ("balanced" is generally taken for granted):

1. You need to know, or be given, the reactants and products actually observed under the given reaction conditions.

> At first, you will be given the names and formulas; gradually, you will acquire criteria for deciding what the substances are. Example: CO_2 and H_2O are the usual products in the "combustion" reactions of carbon- and hydrogen-containing compounds with oxygen or air.

> You will need to use the molecular formula when a substance is known to be molecular; otherwise, the simplest formula. (The maximum information possible is always incorporated in an equation.)

2. Show the physical states of all reactants and products.

> These must be the states you would observe under the given reaction conditions. Example: If water is a product in a reaction occurring at room temperature, you would expect it to be a liquid.

3. Follow the various conventions such as writing the reactants (the substances consumed or disappearing during a reaction) to the left, products to the right, and separated by an arrow ($\rightarrow$), using the simplest ratio of whole numbers of formula units needed to balance the equation. (Balancing requires choosing the relative numbers of formula units so as to conserve atoms — not rewriting the formulas themselves. Example:

> The formation of liquid water from hydrogen and oxygen gases, under almost all conditions, can be represented by the equation

$$2\,H_2\,(g) + O_2\,(g) \rightarrow 2\,H_2O(l)$$

> but *not* by the following "equation,"

$$H_2\,(g) + O(g) \rightarrow H_2O(l)$$

> where the formula of oxygen has been manipulated so as to achieve a balance. The formulas must correspond to reality: oxygen is known to be diatomic.)

Finally, be aware of some of the limitations of chemical formulas and equations. Beyond specifying the physical state, a formula in itself, or an equation, does not directly say anything about the conditions needed for the reaction to occur, or indeed whether the reaction can occur. An equation says nothing about reaction speed; nothing about the extent to which a given reactant is consumed; and nothing about how the reaction actually occurs among the individual atoms and molecules. All of these questions need to be answered by observations that are not represented by an equation.

SELF-TEST

True or False

1. Since three-fourths of the atoms in a sample of $ScCl_3$ are chlorine atoms, 75% of the weight of the sample is due to chlorine. ( )

2. The molecular formula for hydrogen peroxide, H_2O_2, indicates that one molecule of hydrogen (H_2) is bonded to one molecule of oxygen (O_2). ( )

3. In 50 grams of calcium carbonate, $CaCO_3$ (FW = 10)-, there are: 3×10^{23} Ca atoms, 0.5 mole of C, and 3/2 gram atomic weights of oxygen. ( )

4. The materials which are consumed, disappear, during a ( )
chemical reaction are called products.

5. The calculation of theoretical yield is based on the assump- ( )
tion that all of the limiting reactant is consumed according to the
equation for the reaction.

6. All of the following may be reasons why the actual product ( )
yield in a reaction is usually less than 100%: separation and purification results in losses; competing reactions form other products
instead; the reaction hasn't stopped yet; not all the reactants are
converted to desired product, even when the reaction has ceased.

7. The chemical mole is defined as being Avogadro's number of ( )
formula units (molecules, ions, atoms, electrons . . .).

8. In an ordinary chemical reaction, the number of moles of ( )
reactants always equals the number of moles of products.

Multiple Choice

9. The formula weight of lead iodide, PbI_2, is: ( )
   (a) $82 + 2(53)$ (b) $2(82 + 53)$
   (c) $207 + 2(127)$ (d) $207 + 127$

10. In analyzing a compound for carbon, the compound is ( )
burned in air and the masses of products determined. What
assumption do we make?
   (a) all the oxygen in the air is converted to water, H_2O
   (b) all the carbon is converted to carbon dioxide, CO_2
   (c) equal numbers of moles of CO_2 and H_2O are produced
   (d) all the oxygen in the H_2O produced comes from the
       compound

11. An empirical or simplest formula of a substance always shows ( )
   (a) the element(s) present and the simplest ratio of whole
       numbers of atoms
   (b) the actual numbers of atoms combined in a molecule of
       the substance
   (c) the number of molecules in a sample of the substance
   (d) the gram molecular weight of the substance

12. What *minimum* information would be sufficient for ( )
determining the simplest formula for a compound?
    (a) the elements present in the compound
    (b) the elements in the compound and their atomic weights
    (c) the elements in the compound, their atomic weights, and
        their combining weights in a sample of the compound
    (d) the elements in the compound, their atomic weights,
        their combining weights and weight percentages in a
        sample of the compound

13. In order to determine the composition of a substance, one ( )
might
    (a) determine its physical properties and compare them to
        those of known substances
    (b) convert the substance to one or more substances of
        known composition
    (c) compare the products for each of several reactions to
        those gotten when similarly reacting other, known
        substances
    (d) all of the above

14. Information given by the formula for hydrazine, N_2H_4, ( )
includes all of the following except:
    (a) hydrazine could just as well be represented by the
        formula NH_2
    (b) the per cent by weight that is nitrogen is $(28.0/32.0) \times$
        100%
    (c) one molecule of hydrazine contains six atoms
    (d) one mole of hydrazine weighs 32.0 grams

15. The formula for calcium carbonate ($CaCO_3$, FW = 100) ( )
generally represents all of the following except:
    (a) one formula unit ("molecule") of calcium carbonate
    (b) N formula units of calcium carbonate
    (c) one gram of calcium carbonate
    (d) one hundred grams of calcium carbonate

16. A compound with the simplest formula C_2H_5O has a molecu- ( )
lar weight of 90. The molecular formula for the compound is:
    (a) $C_3H_6O_3$ (b) $C_4H_{26}O$ (c) $C_4H_{10}O_2$ (d) $C_5H_{14}O$

17. Which contains the largest number of molecules? ( )
    (a) 1.0 g CH_4 (MW = 16)   (b) 1.0 g H_2O (MW = 18)
    (c) 1.0 g HNO_3 (MW = 63)   (d) 1.0 g N_2O_4 (MW = 92)

18. In exactly one mole of baking soda, $NaHCO_3$, there is about ( )
how much oxygen?

    (a) 16 g   (b) 24 g   (c) 48 g   (d) 96 g

19. It is estimated that there are 1×10^{21} kg of water in all the ( )
oceans. How many moles of water is this?

    (a) $\dfrac{1 \times 10^{21}}{18}$ (b) $\dfrac{1 \times 10^{21} \times 10^{3}}{18}$

    (c) $1 \times 10^{21} \times 18 \times 10^{3}$ (d) $\dfrac{1 \times 10^{21}}{6 \times 10^{23}}$

20. The molecular weight of a protein that causes food poisoning ( )
is about 900,000. The approximate mass of one molecule of this
protein is:

    (a) 2×10^{-18} g   (b) 1×10^{-6} g
    (c) 9×10^{5} g   (d) some other number

21. One mole of the compound Na_2SO_3 contains: ( )
    (a) one mole of Na^+ ions   (b) one atom of S
    (c) one gram of compound   (d) $3 \times 6.02 \times 10^{23}$ atoms of O

22. A balanced chemical equation shows ( )
    (a) the mole ratio in which substances react
    (b) the direction a chemical system will move in and the
        extent or yield of the reaction
    (c) the speed with which the reaction proceeds
    (d) the individual molecular steps by which the reaction
        occurs

23. To write a chemical equation for a given reaction, you would ( )
need to know at least:
    (a) the masses of the reactants and products in the given
        reaction
    (b) the mole ratios of all reactants and products in the
        reaction
    (c) the formulas of all reactants and products in the reaction
    (d) all of the above

24. Ammonia (NH_3) and oxygen (O_2) can be made to react to ( )
form only nitrogen and water. The number of moles of oxygen
consumed for each mole of nitrogen formed is:

    (a) 1.5   (b) 0.50   (c) 3.0   (d) 2.0

25. Which equation most completely represents the following ( )
reaction? On being heated, gaseous ammonia (NH_3) decomposes to
form gaseous nitrogen (N_2) and hydrogen (H_2).

(a) $NH_3(g) \rightarrow N_2(g) + H_2(g)$
(b) $3 H_2(g) + N_2(g) \rightarrow 2 NH_3(g)$
(c) $2 NH_3(g) \rightarrow 6 H(g) + 2 N(g)$
(d) $2 NH_3(g) \rightarrow 3 H_2(g) + N_2(g)$

26. When balanced, the following equation for the combustion of ( )
octane has which set of coefficients?

$$\_\_ C_8H_{18}(l) + \_\_ O_2(g) \rightarrow \_\_ CO_2(g) + \_\_ H_2O(g)$$

(a) 1,25,8,18 (b) 1,25/2,16,18 (c) 1,25,8,9 (d) 2,25,16,18

27. If 4.5 moles of NO_2 are allowed to react with 3.0 moles of ( )
H_2O according to the equation:

$$3 NO_2(g) + H_2O(l) \rightarrow 2 HNO_3(l) + NO(g)$$

the theoretical yield of HNO_3, in moles, would be:

(a) 4.5   (b) 3/2 × 4.5   (c) 2.0   (d) 2/3 × 4.5

28. Baking soda ($NaHCO_3$) and hydrochloric acid (HCl) react ( )
according to the equation: $HCO_3^-(aq) + H^+(aq) \rightarrow H_2O(l) + CO_2(g)$.
At least how many moles of $NaHCO_3$ are required for the formation
of 2.5 moles of CO_2?

(a) 1.0   (b) 2.5   (c) 5.0   (d) some other number

29. How many grams of baking soda (FW = 84.0) would be ( )
consumed in forming 10.0 grams of carbon dioxide (MW = 44)
according to the equation given in (28)?

(a) $\dfrac{10.0}{44}$ (b) $\dfrac{44 \times 84.0}{10.0}$ (c) $\dfrac{10.0 \times 84.0}{44}$ (d) $\dfrac{10.0 \times 61.0}{44}$

30. When balanced using the smallest whole numbers possible, ( )
the coefficient for Cl^- is:

$$\_\_ Cl_2(aq) + \_\_ OH^-(aq) \rightarrow \_\_ Cl^-(aq) + \_\_ ClO_3^-(aq) + \_\_ H_2O(l)$$

(a) 1   (b) 3   (c) 4   (d) 5

31. When a certain compound of nitrogen and hydrogen reacts ( )
with oxygen, nitric oxide (NO) and water are formed. Under reaction
conditions, all four substances are gases and are found to react in a
volume ratio of 4:5:4:6, respectively. What is the simplest formula
for the nitrogen-hydrogen compound?

(a) NH_2 (b) N_2H_4 (c) HN_3 (d) NH_3

32. In carrying out the reaction, ( )

$$2\,NaHCO_3\,(s) \rightarrow Na_2CO_3\,(s) + H_2O(g) + CO_2\,(g)$$

a student obtains an 80% yield. How many moles of $NaHCO_3$ must she have started with if her yield was 1.6 moles of Na_2CO_3?
    (a) 4.0   (b) 3.2   (c) 2.6   (d) 2.0

33. When an excess of one reactant is used: ( )
    (a) more product may form than with no excess
    (b) reaction may proceed at a faster rate
    (c) less limiting reactant may remain unconsumed
    (d) all of the above

SELF-TEST ANSWERS

1. F (Atom for atom, Sc is heavier than Cl; actual % Cl = 70.)
2. F (The formula denotes only that there are two H atoms and two O atoms in a molecule.)
3. T
4. F (Reactants.)
5. T
6. T (These answers are based on material of Chapters 1, 13, and 14 as well as on actual laboratory experience.)
7. T
8. F
9. c (Simply add the atomic weights for all atoms: see Chapter 2.)
10. b
11. a
12. c (Note that combining weights and weight percentages provide the same information.)
13. d (Any one or all of these might work.)
14. a (NH_2 would not give the correct number of atoms in a molecule of hydrazine, as indicated by the molecular formula N_2H_4.)
15. c
16. c (All of these formulas correspond to a MW of 90; but only (c) has the atom ratio given also by the simplest formula.)
17. a (The largest fraction of a mole, 1.0/16.)
18. c
19. b (Don't forget to convert to grams!)
20. a (Mass per molecule = 900,000 g/6 × 10^{23} molecules.)
21. d
22. a

23. c (Note that the information in (a) and (b) can be gotten from this choice, once the equation is balanced.)
24. a (You need to know that nitrogen is N_2 and to balance the equation.)
25. d
26. d
27. d (Note that NO_2 is the limiting reactant; 4.5 moles NO_2 $\times$ $\dfrac{2 \text{ moles } HNO_3}{3 \text{ moles } NO_2}$.)
28. b (Each mole of HCO_3^- reacting requires the use of one mole of $NaHCO_3$.)
29. c
30. d (Difficult? A systematic approach to balancing such an equation is given in Chapter 20. The respective coefficients are: 3, 6, 5, 1, 3.)
31. d (Recall Avogadro's hypothesis – Chapter 2: Volume ratios are equivalent to mole ratios for gases reacting under a given set of conditions. The equation:

$$4 NH_3(g) + 5 O_2(g) \rightarrow 4 NO(g) + 6 H_2O(g).)$$

32. a (For 80% yield, every 2 moles $NaHCO_3$ gives 0.8 mole Na_2CO_3.)
33. d (Choices (a) and (c) are discussed in Chapter 13; (b), in Chapter 14.)

SELECTED READINGS

Problem solving requires practice. For a good selection of exercises use any one of the problem or programmed manuals listed in the Preface to this guide. (Also, work as many of the problems in the text as you can.)

Copley, G. N., Linear Algebra of Chemical Formulas and Equations, *Chemistry* (October) 1968), pp. 22-27.
   A general method for balancing equations by the use of determinants. Recommended only for the mathematically inclined, this is one of a series of articles.
Kieffer, W. F., *The Mole Concept in Chemistry*, New York, Reinhold, 1963.
   A few exercises, most of them worked out, are provided in this general discussion of stoichiometry. Will be useful later in the course too.

4 THERMOCHEMISTRY

QUESTIONS TO GUIDE YOUR STUDY

1. How many chemical reactions going on in the world around us can you think of in which the energy change plays an important role? (What "drives" an automobile engine; a mountain climber?)

2. What kinds of energy may be transferred during chemical reactions?

3. What is the source of the energy involved in a reaction? What happens to it? Can this energy be rationalized in terms of what happens to the atoms?

4. Is the energy transferred during a reaction quantitatively related to the reacting masses? (Can you cite a simple example?)

5. How is information concerning energy transfer expressed within the chemical equation? Does your interpretation of the energy transfer depend on how you write the equation for a reaction?

6. Does the energy transfer depend on the conditions under which a reaction is carried out? If so, how?

7. How do you measure the energy transfer for any given reaction? Can it be calculated for a reaction, without ever actually carrying out that particular reaction?

8. What do the thermochemical principles allow you to say about practical problems, such as the relative merits of two different fuels?

9. What are some of the fundamental problems in supplying the energy needed by modern society today and tomorrow? Are there known limitations or perhaps untapped possibilities the chemist can describe?

10. What is *thermal* pollution? What can be done to minimize it?

11.

12.

YOU WILL NEED TO KNOW

Concepts

1. A general notion as to the meaning of energy and the means by which it may be transferred from one object to another — See the Summary, as well as the Readings

Math

1. How to work problems in stoichiometry, and therefore how to write simple balanced equations — Chapter 3

CHAPTER SUMMARY—OBJECTIVES

In every chemical reaction we observe that the mass of the products is equal to the mass of the reactants. We account for mass conservation in writing equations that are "balanced." We likewise have a bookkeeping system for energy since it too can always be accounted for. In a "thermo-chemical" equation we note how much energy is involved and show whether it is entering or leaving the system (the reaction mixture). The stoichiometry is as before — the equation is taken to represent amounts of substances in units of mole, with the accompanying energy change having the numerical value written into the equation or alongside it.

The usual way in which we measure the energies associated with chemical reactions is to measure a property of the surroundings, such as temperature, that shows how much energy has left the reaction mixture and entered the surroundings, or left the surroundings and entered the system. In all reactions, the energy is thought of as either being stored in the atoms and molecules of the system (the energy of the system increases, the sign of the energy change is positive) or being taken out of storage and passed on to the surroundings (where the energy is stored, again, in the constituent particles). What happens during a chemical reaction if the system is insulated from surrounding matter? The energy of reaction is involved in a change from one kind of storage to another. (Example: If hydrogen and oxygen explosively react to form water inside a sealed and insulated "bomb," the energy that would otherwise have been given off to the surroundings ends up being stored in the products, raising their temperature and increasing the vigor of their molecular motions.)

The transfer of energy of most concern to the chemist is that called heat, or heat flow or thermal energy. Since most reactions occur open to the atmosphere, as in test tubes, we are interested in the thermal energy transferred at constant pressure. This is called the enthalpy change, ΔH. For most purposes, even if the pressure does change during a reaction, the heat flow is equal to ΔH.

How do we associate energy with a molecule? In what ways can energy be stored; whence does it come? For the very fact that molecules are constantly in motion, they possess energy; the more energy, the more vigorous the motion. (What kind of evidence is there for molecular motion?) More about kinetic energy in the next chapter. And several kinds of motion may be possible for a molecule, atom, or ion: vibrating, rotating, or just moving headlong. And there's the stored "chemical energy" associated with the bonds between atoms: a result of mutual attractions, and motions, between electrons and nuclei. Differences from one molecule to the next kind of molecule in the types and numbers of these energy storing mechanisms give rise to differences in the energy transfers observed. For example, a lot more energy is normally stored in a chemical bond between the atoms in a molecule than in the motions of the molecule itself. Or again, much more energy is involved in nuclear changes than in the ordinary bond-making and bond-breaking of chemical reactions.

The fact that we can keep track of energy transfers, that energy always seems to be conserved in chemical reactions, means that we have the power of prediction. We can often predict the energy effect associated with a given reaction, even without first carrying out the reaction. And this is really the chief concern of this chapter — that you be able to calculate the amount of heat produced by or required for any given reaction.

But perhaps there is a more important concern for the material discussed in this chapter: that of objectively looking at our energy resources on this planet, their uses and abuses, the laws and limitations energy changes seem to follow. Just one such observation, with vast implications extending beyond chemistry, is that, as far as we can tell, heat can never be completely changed into work; there is always some unused and unusable heat left over. (In Chapter 12, we will see a rationale for the "inevitable" waste of energy and in terms of the behavior of particles.) In essence, the "first law" of thermodynamics says that energy is not really consumed at all; it can only be changed in form; for example, from chemical energy into heat. The "second law" might be taken to mean that energy cannot be completely recycled; more and more, energy becomes unavailable as wasteful heat in the surroundings. How, then, do we make the best of what we've got?

Objectives

After studying this chapter, you should be able to
1. use calorimetric data to calculate heats of reaction;

2. calculate ΔH for any reaction, given heats of formation or heats of bond formation; and in general apply Hess's law;

3. distinguish between the various types of heats of reaction (in particular, recognize the corresponding chemical equation);

4. interpret the sign of the enthalpy change; interpret and construct enthalpy diagrams;

5. begin to interpret energy changes in terms of molecular properties;

6. interpret fuel or heating values of foods and fuels;

7. outline trends in energy sources and uses.

SELF-TEST

True or False

1. In carrying out a reaction in a test tube, a student observes ( )
that the test tube becomes cold. He should call the reaction exothermic.

2. Given the thermochemical equation: ( )

$$UF_6(l) \rightarrow UF_6(g), \Delta H = +7.2 \text{ kcal},$$

one can be sure that at least 7.2 kcal of heat must be produced or evolved when one mole of liquid UF_6 is evaporated.

3. Another way of writing the thermochemical equation of ( )
question (2) would be: $UF_6(l) + 7.2 \text{ kcal} \rightarrow UF_6(g)$.

4. You would expect that the heat produced in the following ( )
reactions would have one and the same numerical value:

$$H_2(g) + \tfrac{1}{2} O_2(g) \rightarrow H_2O(l); \quad 2 H_2(g) + O_2(g) \rightarrow 2 H_2O(l)$$

5. The difference in enthalpy between one mole of Cl_2 and two ( )
moles of atomic chlorine, Cl, both at one atmosphere pressure and $25°C$ is equal in magnitude to the heat of bond formation for one mole of Cl_2.

6. Considering the example reaction: $C(s) + O_2(g) \rightarrow CO_2(g)$, ( )
one can always say that the heat of combustion for any substance is the same thing as the so-called heat of formation of that substance.

7. One would expect that ΔH for the following reaction would ( )
be smaller (i.e., less negative) than the molar heat of formation of liquid water:

$$2 H(g) + O(g) \rightarrow H_2O(l)$$

8. If a system is carried through a series of steps, each of which   (   )
involves an energy transfer into or out of the system, but finally ends
up being identical in every way to the initial system, the total thermal
energy transfer must be zero.

9. Reactions which tend to occur of their own accord, proceed-   (   )
ing "naturally" or spontaneously, usually involve a decrease in
enthalpy.

Multiple Choice

10. The largest amount of energy is stored in which one of the   (   )
following systems?
   (a) 1.0 mole $H_2O(l)$ at $100°C$
   (b) 1.0 mole $H_2O(l)$ at $0°C$
   (c) 1.0 mole $H_2O(g)$ at $100°C$ and 1.0 atmosphere
   (d) 1.0 mole $H_2O(g)$ at $100°C$ and 1.0 atmosphere and 1.0
        mole $H_2O(l)$ at $100°C$ have the same amount of stored
        energy

11. The regulation of body temperature by perspiration can be   (   )
accounted for, at least in part, by the fact that
   (a) liquid water has a negative heat of formation
   (b) liquid water is a good insulator
   (c) the phase change, liquid-to-gas, is endothermic
   (d) water absorbs and releases heat more rapidly than most
        substances

12. The molar heat of combustion of methane, CH_4, is reported   (   )
as -213 kcal. The corresponding chemical equation is:
   (a) $CH_4(g) + \frac{3}{2} O_2(g) \rightarrow CO(g) + 2 H_2O(l)$ $\Delta H = -213$ kcal
   (b) $CH_4(g) + 2 O_2(g) \rightarrow CO_2(g) + 2 H_2O(l)$ $\Delta H = -213$ kcal
   (c) $C(s) + 2 H_2(g) \rightarrow CH_4(g)$ $\Delta H = -213$ kcal
   (d) $C(g) + 4 H(g) \rightarrow CH_4(g)$ $\Delta H = -213$ kcal

13. One can often confidently calculate the enthalpy change for   (   )
a reaction before carrying out the reaction because
   (a) all heats of reaction have already been measured and
        tabulated
   (b) heats of formation for all known compounds have been
        measured
   (c) the given reaction, and its enthalpy change, may be
        related algebraically to other reactions already carried out
   (d) since enthalpy change depends on amount of substance
        reacting, one can always adjust the amount of materials
        so as to observe an enthalpy change of any value

14. For the reaction: $2 HCl(g) \rightarrow H_2(g) + Cl_2(g)$, $\Delta H = +44.2$ (   )
kcal. This means that:
  (a) if the given reaction is to be carried out at constant temperature, then the reaction mixture must be heated
  (b) the chemical bonds in the products are weaker than those in the reactants
  (c) $HCl(g)$ has a negative heat of formation
  (d) all of the above

15. When water evaporates at constant pressure, the *sign* of the (   )
heat effect
  (a) is negative
  (b) is positive
  (c) depends on the temperature
  (d) depends on the container volume

16. Given $\Delta H = -143.8$ kcal for the reaction: $Mg(s) + \frac{1}{2} O_2(g) \rightarrow$ (   )
$MgO(s)$, what would you expect to happen if the reaction were allowed to proceed at constant pressure in such a way that no heat transfer could take place between the reaction mixture and the surroundings?
  (a) no reaction could occur
  (b) the temperature of the reaction mixture would increase
  (c) the temperature of the reaction mixture would decrease
  (d) insufficient information is given

17. When 1.00 gram of ammonia, NH_3 (FW = 17.0), is produced (   )
from N_2 and H_2 at a constant temperature (25°C) and pressure (1 atm), 648 calories are produced. The molar heat of formation of ammonia, in kilocalories, is
  (a) -0.648 (17.0)   (b) 17.0/648
  (c) $-0.648/17.0$ (d) $+0.648$ (17.0)

18. Which one of the following reactions would you expect to be (   )
the source of the largest amount of heat?
  (a) $CH_4(l) + 2 O_2(g) \rightarrow CO_2(g) + 2 H_2O(g)$
  (b) $CH_4(g) + 2 O_2(g) \rightarrow CO_2(g) + 2 H_2O(g)$
  (c) $CH_4(g) + 2 O_2(g) \rightarrow CO_2(g) + 2 H_2O(l)$
  (d) $CH_4(g) + \frac{3}{2} O_2(g) \rightarrow CO(g) + 2 H_2O(l)$

19. In what order would you arrange the following reactions so ( )
that the enthalpy change increased in that order?

<blockquote>
A. $^{238}U + n \rightarrow ^{239}U$
B. $H_2O(l) \rightarrow H_2O(s)$
C. $C(coal) + O_2(g) \rightarrow CO_2(g)$
</blockquote>

(a) ABC   (b) BCA   (c) CAB   (d) ACB

20. Given the heats of combustion for diamond (-94.50 kcal) ( )
and graphite (-94.05 kcal), both composed of pure carbon, would
you expect the formation of diamond from graphite to be
endothermic or exothermic?
(a) endothermic   (b) exothermic
(c) depends on the temperature   (d) insufficient information

21. Hydrogen, H_2, has been proposed as a fuel of the future. ( )
Knowing that very little hydrogen exists in this elemental state, you
would expect that
(a) hydrogen could play no role in the transfer of energy
from one system to another
(b) elemental hydrogen might well supplant all other fuels
(c) energy received from another source might be stored in
H_2 and later released during the combustion of the
hydrogen
(d) all of the above

SELF-TEST ANSWERS

1. F (To restore the temperature to its initial value, heat must
eventually be added to the test tube.)
2. F
3. T
4. F (The first heat effect would be that for forming one mole of
water; the second, twice as much heat for twice the amount of
water. This is how the equation stoichiometry is interpreted.)
5. T
6. F (Solid carbon is undergoing combustion; gaseous carbon dioxide
is being formed from the elements.)
7. F (The given reaction can be considered as the sum of the two
reactions: $H_2(g) + \frac{1}{2} O_2(g) \rightarrow H_2O(l)$; $2 H(g) + O(g) \rightarrow H_2(g) + \frac{1}{2} O_2(g)$. The second step releases additional energy, making the
overall ΔH more negative.)

8. F   (Q could be practically anything, provided other kinds of energy transfer were involved.)

9. T   (You probably expect this from common experience: heat is generally produced by reactions. For a closer look at spontaneous reactions, see Chapter 12.)

10. c   (ΔH is positive for $H_2O(l) \rightarrow H_2O(g)$; additional energy is stored in the gas.)

11. c   (Evaporation requires heat; heat is removed from the hot body.)

12. b   (Note that CO_2 is generally formed, and not CO, in the reaction chemists call combustion — unless otherwise noted.)

13. c

14. d

15. b   (Choice c, for example, is ruled out since ΔH doesn't change much with temperature.)

16. b

17. a   (Check your units.)

18. c   (The energy released increases from a to b to c; as well as from d to c, since burning CO to form CO_2 would release the larger amount of energy associated with c. Do you also see the rationale for always specifying the states?)

19. b

20. a   ($\Delta H = +0.45$ kcal/mole. This is small compared to most heats of reaction for chemical changes. So why were diamonds only recently synthesized?)

21. c   (Readings.)

SELECTED READINGS

The Biosphere, San Francisco, W. H. Freeman, 1970 (the September 1970 issue of *Scientific American*).
    See particularly the chapters on the energy cycles of the earth and on human energy production.

Daniels, F., *Direct Use of the Sun's Energy*, New Haven, Yale, 1964.
    A pioneering and comprehensive work on solar energy; illustrated, almost "how to do it."

Faraday, M., *The Chemical History of a Candle*, New York, Viking Press, 1960.
    A delightful exploration of many aspects of combustion; dates back to 1860.

Holdren, J. and P. Herrera, *Energy*, New York, Sierra Club, 1971.
    A readable, apparently unbiased account of present and likely future sources, uses and abuses of energy. Includes "case studies" of citizen action groups.

Hydrogen: Likely Fuel of the Future, C & EN (June 26, 1972), pp. 14-17.
    The first in a three-part series describing "a new energy system" employing hydrogen as a carrier (not a primary source) of energy; sources of hydrogen; implications of large scale use.

Mahan, B. H., *Elementary Chemical Thermodynamics*, New York, W. A. Benjamin, 1963.
   A good introduction to classical thermodynamics, going somewhat farther and deeper into the subject than this text. (Also see the paperback below and other references listed for Chapter 12.)
Pimentel, G. C. and R. D. Spratley, *Understanding Chemical Thermodynamics,* San Francisco, Holden-Day, 1969.
   With humor and lots of commonsense illustrations and applications, the authors introduce you to most of the uses and meanings you are likely to want, in this course and elsewhere.
Sandfort, J. F., *Heat Engines: Thermodynamics in Theory and Practice*, Garden City, New York, Doubleday-Anchor, 1962.
   A very easily read introduction to engineering thermodynamics and its history.
Strong, L. E. and W. J. Stratton, *Chemical Energy*, New York, Reinhold, 1965.
   A very carefully developed discussion – though not easy reading – on the same level as the text, with an emphasis on interpreting thermodynamics in terms of molecular properties (structure and bonding; disorder).

For further practice in the calculations of thermochemistry, see Barrow, and the other problem manuals listed in the preface.

5 PHYSICAL BEHAVIOR OF GASES

QUESTIONS TO GUIDE YOUR STUDY

1. What materials can you think of that usually exist as gases? What properties do they share?

2. What conditions favor the existence of a substance as a gas rather than as a liquid or a solid?

3. How do you measure properties of gases such as temperature and pressure? How would you weigh a sample of gas?

4. Is there a simple relationship among the properties of a gas that generally holds for all gases? Can you rationalize such a relationship in terms of the behavior of molecules? (For example: how does the behavior of molecules explain the relation between temperature and pressure for the air inside a tire?)

5. How do mixtures of gases behave? How is their behavior related to that of a pure gaseous substance?

6. Can the quantities of gases participating in a chemical reaction be simply expressed in terms of weights; moles; volumes?

7. How does the volume of gaseous reactant or product depend on reaction conditions?

8. What are some of the practical applications (as well as support for other chemical principles) of our knowledge of gaseous behavior?

9. What experimental support is there for our ideas about the nature of the molecules in a gas?

10. How do you account for the differences in properties among gases and between gases and liquids and solids?

11.

12.

YOU WILL NEED TO KNOW

Concepts

1. Meaning of molecular weight (and how to calculate it from a formula) — Chapter 2

Math

1. How to solve first and second order equations for any one variable

Example: Rewrite the equation $pV = \dfrac{gRT}{M}$ in the form $M = \ldots$

Example: Solve the equation $\frac{1}{3} Mu^2 = RT$ for u.

2. How to find the square root of a number (most simply done by using a slide rule or a table of logarithms) — see readings listed in the Preface.

CHAPTER SUMMARY—OBJECTIVES

The physical behavior of gases is described concisely by the ideal gas law

$$pV = nRT,$$

which forms the central theme of this chapter. This equation tells us how the four experimental variables, pressure (p), volume (V), number of moles (n), and absolute temperature (T) are related to one another. We see, for example, that for a given sample of gas at a fixed temperature (i.e., n and T constant):

$$pV = \text{constant, or } p = \text{constant}/V \quad \text{(Boyle's Law).}$$

Again, we are reminded that for a given sample of gas at a fixed pressure (i.e., p and n constant):

$$V = \text{constant} \times T \quad \text{(Law of Charles and Gay-Lussac).}$$

In still another case, we see that when temperature and pressure are held constant:

$$V = \text{constant} \times n,$$

which implies Avogadro's Law (equal volumes of all gases at the same temperature and pressure contain equal numbers of molecules).

Frequently, we use the ideal gas law to calculate one variable (p, V, n or T) given the values of the other three. In order to carry out such calculations, we must know the magnitude of the gas law constant, R, in the proper units. For our purposes in this chapter, it is most convenient to express R in lit atm/mole °K:

$$R = 0.0821 \text{ lit atm/mole } °K.$$

From time to time, particularly in later chapters of this text, we will use R in other units:

$$R = 1.99 \text{ cal/mole } °K = 8.31 \times 10^7 \text{ ergs/mole } °K.$$

Note that R always has the units of energy/mole °K (a liter atmosphere represents the work done when a piston sweeps through a volume of one liter against a constant pressure of one atmosphere).

For certain applications of the ideal gas law, it is convenient to substitute for the number of moles, n, its equivalent in grams, i.e., $n = g/M$, where g is the number of grams of the gas and M its gram molecular weight. The resulting equation

$$pV = \frac{gRT}{M}$$

can be used to determine the molecular weight of a gas from measured values of p, V, g and T. Furthermore, recognizing that the density is mass divided by volume, $D = g/V$, we obtain the relation

$$D = pM/RT$$

which tells us, among other things, that the densities of different gases at the same temperature and pressure are in the same ratio as their molecular weights.

The ideal gas law can be applied to gas mixtures as well as to pure gases. We can, for example, write: $p_a V = n_a RT$, where p_a is the partial pressure and n_a the number of moles of gas A in the mixture. By combining similar expressions for each gas in the mixture, it is possible to obtain Dalton's Law

(Equation 5.8, p. 105, text). This law is particularly useful in making calculations involving "wet" gases. Such mixtures, in which water vapor is one component, are commonly formed in the laboratory when gases are collected over water.

The validity of the ideal gas law is confirmed by experimental measurements which require no assumptions about the behavior of gas molecules. However, by making some rather simple assumptions concerning molecular motion, embodied in what is known as the kinetic molecular theory, it is possible to derive the law from first principles. This exercise in logic was one of the great triumphs of 19th century science; indeed, it provided the first really convincing evidence for the existence of molecules and atoms. A key postulate of kinetic theory is that, at a given temperature, molecules of all gases have the same kinetic energy of translation. Specifically

$$\epsilon = mu^2/2 = \text{constant} \times T$$

where m is the mass and u the average velocity of a molecule. Starting with this postulate, it is possible to derive equations for the relative rates of effusion of different molecules (Graham's Law, Equation 5.16, p. 112, text) or the average velocity of a particular molecule at a given temperature (Equation 5.17, p. 114). You should keep in mind that u in these equations is an *average* velocity; at any given instant, virtually all of the molecules are moving at velocities either greater or smaller than u.

The simple kinetic molecular model of gases ignores interactions between molecules and assumes their volume to be negligible in comparison to that of their container. At low pressures and high temperatures, where the molecules are far apart, these approximations are justified and we find that the ideal gas law describes very well the behavior of real gases. However, at high pressures and low temperatures, intermolecular forces and molecular volumes become significant and real gases deviate considerably from ideal behavior. Indeed, if the molecules of a gas approach each other closely enough, condensation to a liquid or solid occurs, in which case the ideal gas law is inapplicable.

Objectives

You will find that the material in this chapter can best be mastered by working problems. Specifically, you should be able to use the ideal gas law to

1. derive the relation between two (e.g., p and V) or three (e.g., p, V, T) variables, other factors (e.g., n and T or n alone) being constant;

2. apply such derived relations to calculate the final state of a gas given the initial conditions;

3. evaluate one variable (p, V, n or T) given the other three;

4. determine the density of a particular gas as a function of T and p;

5. calculate the molecular weight of a gas from its measured density at a known p and T;

6. make calculations involving the volumes of gases taking part in reactions.

You should also know how to apply

7. Dalton's Law to obtain partial pressures of gases in mixtures;

8. Graham's Law to obtain relative rates of effusion;

9. Equation 5.17, p. 114, to calculate the average velocity of a particular gas molecule at a given T; to find typical values at or near room temperature.

In addition, you should be able to interpret

10. absolute zero and temperature itself in terms of molecular behavior;

11. deviations from the ideal gas law in terms of molecular behavior.

And, you should be able to

12. describe the means by which gas pressures are usually measured.

SELF-TEST

True or False

1. To prepare pure nitrogen gas from a sample of air, one would probably employ fractional distillation.　　　　　　　　　　( )

2. The molecules of a sample of any gas, under most conditions, can be characterized as being separated by relatively large distances.　( )

3. At a temperature of 20°C, about 3.4 grams of carbon dioxide gas will dissolve in a liter of water under a pressure of 1.0 atmosphere. At a higher temperature, but at the same pressure, one would expect that more carbon dioxide will dissolve.　　　　　　　( )

4. Under some conditions, pV = nRT may not accurately describe the behavior of a gas. This can often be accounted for by assuming that attractive forces exist between molecules.　　　　　( )

5. In a mixture of two gases, with a total pressure of 760 mm Hg, the partial pressure of chlorine gas is 380 mm Hg. This means that half of the molecules in the sample are chlorine.　　　　　　( )

6. The average kinetic energy of an oxygen molecule and the average kinetic energy of a hydrogen molecule, both gases at the same temperature, are in the same ratio as their molecular weights.　( )

7. The average speed of a gas molecule depends only on the absolute temperature.　　　　　　　　　　　　　　　　　( )

8. Two separate samples of the same gaseous substance at the   (   )
same pressure would have densities in the same ratio as their absolute
temperatures.

9. The fact that a sample of gas would not have zero volume at   (   )
the absolute zero of temperature is a consequence of the fact that
absolute zero cannot be reached.

Multiple Choice

10. The chemical analysis of a mixture of gases is most likely to   (   )
involve
   (a) density measurements   (b) mass spectroscopy
   (c) boiling point measurements   (d) fractional crystallization

11. Which one of the following substances would you expect to   (   )
normally exist as a gas at room temperature?
   (a) CH_4 (b) C_2H_6 (c) C_3H_8 (d) C_3H_7OH

12. The volume of a mole of gas is   (   )
   (a) 22.4 liters
   (b) directly proportional to pressure and absolute tempera-
       ture
   (c) directly proportional to pressure, inversely proportional
       to absolute temperature
   (d) inversely proportional to pressure, directly proportional
       to absolute temperature

13. The inflation of an automobile tire to a pressure of "20   (   )
pounds" means:
   (a) the pressure exerted by the air in the tire is 20 lb/in^2
   (b) the weight of the air in the tire is 20 pounds
   (c) the pressure of the air in the tire is 20 lb/in^2 higher than
       the air pressure outside the tire
   (d) the pressure of the air inside the tire is about 5 lb/in^2

14. A certain mountain rises to 14,100 feet above sea level. The   (   )
pressure at the top is about 17.7 inches (of mercury). If you blew up
a balloon at sea level, where the pressure happened to be 29.7 inches,
and carried it to the top of the mountain, by what factor would its
volume change?
   (a) there would be no change   (b) 29.7 − 17.7
   (c) 29.7/17.7   (d) 17.7/29.7

15. When equal numbers of moles of two gases at the same ( )
temperature are mixed in a container, the pressure of the gaseous
mixture is
   (a) given by Gay-Lussac's law of combining volumes
   (b) the product of the pressures each gas would have if alone
   (c) the difference of the pressures each gas would have if
       alone in the container
   (d) the sum of the pressures each gas would have if alone

16. A gaseous substance is known which can be decomposed to ( )
give only the elements phosphorus and hydrogen. When all three
substances are gases at a convenient temperature and pressure, it is
found that four volumes of the compound give one volume of
phosphorus and six volumes of hydrogen. The simplest interpretation
is that phosphorus gas is
   (a) P   (b) P_2 (c) P_3 (d) P_4

17. An equation representing a reaction which is consistent with ( )
the data of question 16 would be:
   (a) $4\ PH_2(g) \rightarrow 2\ P_2(g) + 4\ H_2(g)$
   (b) $4\ PH_3(g) \rightarrow P_4(g) + 6\ H_2(g)$
   (c) $2\ PH_3(g) \rightarrow 2\ P(g) + 3\ H_2(g)$
   (d) some other equation

18. Two flasks of equal volume are filled with different gases, A ( )
and B, at the same temperature and pressure. The weight of gas A is
0.34 gram while that of gas B is 0.48 gram. It is known that gas B is
ozone, O_3, and that gas A is one of the following. Which of the
following is most likely to be gas A?
   (a) O_2 (b) H_2S (c) SO_2 (d) cannot say

19. Which one of the following statements about the gases A and ( )
B of question 18 is false?
   (a) the numbers of molecules of A and B are equal
   (b) the average speeds of molecules A and B are the same
   (c) the masses of individual molecules of A and B compare in
       the same way as the masses of the samples
   (d) the average translational energies of molecules of A and B
       are the same

20. The fact that the ideal gas law only approximately describes   (   ) the behavior of a gas can be partly explained by the idea that
    (a) R is not really a constant
    (b) gas molecules really do have zero volume
    (c) the kinetic energy of gas molecules is not really directly proportional to the absolute temperature
    (d) gas molecules really do interact with each other

21. Real gases behave most nearly like the ideal gas law says they   (   ) do at
    (a) high temperatures, low pressures
    (b) low temperatures, high pressures
    (c) high temperatures, high pressures
    (d) low temperatures, low pressures

22. The van der Waals equation, $p = RT/(V - b) - a/V^2$,   (   ) incorporates the following correction(s) to the ideal gas law in order to account for the properties of real gases:
    (a) the possibility of chemical reaction between molecules
    (b) the finite volume of molecules
    (c) the quantum behavior of molecules
    (d) average kinetic energy is inversely proportional to temperature

23. To convert a sample of air into a liquid, you would probably   (   ) have to
    (a) increase the temperature and pressure of the sample
    (b) decrease the temperature and increase the pressure of the sample
    (c) cool it to $0°K$
    (d) the task is an impossible one

24. Your lecturer opens a bottle of hydrogen sulfide gas, H_2S,   (   ) and a bottle of gaseous diethyl ether, $C_4H_{10}O$, at the same time. Both gases are at the same temperature and pressure. Which of the two should you be able to smell first? (Both have characteristic odors.)
    (a) the ether
    (b) H_2S
    (c) both at the same time
    (d) neither gas would escape into the room

SELF-TEST ANSWERS

1. T (The air would be liquefied and then fractionally distilled. See Chapter 1.)
2. T (Large compared to molecular diameters.)
3. F (Think of what would happen if you warmed up a bottle of carbonated water, or a bottle of Coke. More on solubility — Chap. 10.)
4. T
5. T
6. F (They would be the same.)
7. F (Also on molecular weight.)
8. F (Inversely: density = mass/volume = pM/RT, where M = molecular weight. So, d(1)/d(2) = T(2)/T(1) for gases (1) and (2).)
9. F (Any real gas condenses to a liquid well above absolute zero.)
10. b
11. a (Low molecular weight — Chap. 8.)
12. d (The volume would be 22.4 liters only under standard conditions.)
13. c (Now you know!)
14. c (Larger at lower pressure.)
15. d (Dalton's law of partial pressures.)
16. d (See question 17, also based on Avogadro's hypothesis.)
17. b
18. b
19. b
20. d
21. a
22. b
23. b (These are the conditions under which the real gas deviates most from the gas laws.)
24. b (With the lower molecular weight — and given the same average translational energies — the H_2S would have the higher average speed.)

SELECTED READINGS

Conant, J. B., *Science and Common Sense*, New Haven, Yale, 1951.
   A case history approach to science for the layman; particularly interesting and readable are the sections on Boyle's work with gases and Lavoisier's pioneering experiments on combustion.

Hildebrand, J. H., *An Introduction to Molecular Kinetic Theory*, New York, Reinhold, 1963.

A more extensive discussion than the present chapter, with much more on real gases; the theory is extended to describe some of the properties of liquids and solids. The author often injects himself into the discussion.

Neville, R. G., The Discovery of Boyle's Law, 1661-62, *J. Chem. Ed.* (July 1962), pp. 356-359.

Interesting mainly for the quotations from Boyle.

Steinherz, H. A. and P. A. Redhead, Ultrahigh Vacuum, *Scientific American* (March 1962), pp. 78-90.

A very graphic discussion of the instruments for measuring and producing very low pressures ($< 10^{-8}$ mm Hg), as well as of its uses.

6 ELECTRONIC STRUCTURE OF ATOMS

QUESTIONS TO GUIDE YOUR STUDY

1. On the basis of what you have learned so far, what can be said about the properties of specific substances? (For example: at 150°C and one atm., is water a gas, liquid, or solid; is it colorless; what is its density?)

2. Are there regularities in the observed properties of the chemical elements? (For example, in their molecular formulas?)

3. How can the information referred to in (2) be systematized? How might it be used?

4. Are there irregularities in the properties of elements? If so, how are they accounted for?

5. Can properties of substances be explained in terms of the nature of individual atoms and molecules?

6. What direct evidence is there for the idea that atoms are themselves composed of smaller parts? (Review Chapter 2.)

7. What experimental evidence demonstrates that there are regularities in arrangement of electrons in atoms? (Both within a particular kind of atom and for atoms of all the elements.)

8. How do you conveniently represent the structure of an atom and the energy associated with a change in this structure? How can you describe in a meaningful way where electrons are and what they are doing?

9. Are there general, simplifying correlations between electronic structure of atoms and their properties, such as the strength of the forces between atoms in a molecule?

10. Do you suppose that modern quantum theory is the "final word"?

11.

12.

YOU WILL NEED TO KNOW

Concepts

1. General ideas of atomic and nuclear composition (kinds, numbers, and charges associated with nuclear particles; meaning of atomic number; isotope) — Review Chapter 2
2. A general idea of what comprises the electromagnetic spectrum (visible, ultraviolet and infrared light; x-rays) — the names of the regions and their approximate energies or wavelengths or frequencies — See this chapter, as well as any introductory physics textbook

Math

1. No new math — concepts or operations — is required
2. How to compute the charge of a monatomic ion, given the number of electrons and the nuclear charge (or atomic number) — Chapter 2

CHAPTER SUMMARY—OBJECTIVES

The electronic structures of atoms, the major theme of this chapter, have occupied the attention of chemists and physicists for more than a century. As so often happens, theories were developed and modified to explain certain puzzling experimental observations. In particular, it was found that light emitted by "excited" gaseous atoms appears at certain discrete wavelengths. To explain this phenomenon, it was proposed that electrons in atoms are restricted to discrete energy levels whose distance of separation, ΔE, is related to the wavelength, λ, of the emitted light by the Einstein equation:

$$\Delta E = \frac{hc}{\lambda} = \frac{1.99 \times 10^{-8}}{\lambda}; \quad (\Delta E \text{ in ergs, } \lambda \text{ in Å})$$

This postulate of the quantization of electronic energies is basic to modern quantum theory.

In 1913, Niels Bohr derived from first principles an equation for the allowable energy levels in a one-electron atom. This equation was in remarkably good agreement with the observed spectrum of hydrogen. The Bohr model predicted that the electron, in moving about the nucleus, would be restricted to a finite number of spherical orbitals of fixed radius.

Unfortunately, this simple model breaks down for any species with more than one electron. Modern quantum theory tells us that exact positions of electrons cannot be specified. The best we can hope to do is to calculate the probability of finding an electron in a particular region.

Electronic energies can be calculated, at least in principle, from the Schrödinger wave equation. You were probably relieved to find that we did not attempt in this text to solve this formidable mathematical expression for allowed energies. Instead, we discussed briefly a simpler, less general expression, Equation 6.12, corresponding to the so-called "particle in a box" model. This model helps us to understand why small particles confined to very small regions of space do not obey the classical laws of motion with which we are familiar.

To completely characterize an electron in an atom, we indicate its

(1) *principal energy level,* specified by the quantum number n, which is restricted to positive, integral values (n = 1, 2, 3, – – –);

(2) *sublevel,* designated (illogically) by the letters s, p, d, f. Within a principal level of quantum number n, there are n sublevels. For n = 1, we have only the s sublevel; the second principal energy level has two sublevels (2s, 2p), the third, three (3s, 3p, 3d), and so on. This same information may be expressed by assigning each sublevel a quantum number ℓ and stating that ℓ can take on any integral value from 0 to (n – 1);

(3) *orbital* (Within a sublevel of quantum number ℓ, there are $2\ell + 1$ orbitals. An s sublevel (ℓ = 0) has only one orbital; a p sublevel (ℓ = 1) has three, a d sublevel (ℓ = 2) five, and an f sublevel (ℓ = 3) seven. Orbitals are assigned quantum numbers, $m\ell$, which take on all integral values from $-\ell$ to ℓ);

(4) *spin,* which is restricted to two possible values. These may be indicated as "up" and "down" ($\uparrow\downarrow$) or by assigning a quantum number m_s of $+\frac{1}{2}$ or $-\frac{1}{2}$. Experimentally, we find that no two electrons in an atom can have the same set of four quantum numbers (Pauli exclusion principle); it follows (why?) that a given orbital can contain no more than two electrons.

For many purposes, it is sufficient to describe the electronic structure of an atom by quoting its *electron configuration,* which tells us how many electrons are located in each sublevel. Thus, the electronic configuration for nitrogen (AN = 7), $1s^2 2s^2 2p^3$, tells us that there are 2 electrons in the 1s sublevel, 2 in the 2s, and 3 in the 2p. Electron configurations can be deduced by knowing the order in which sublevels are filled (p. 143, text) and recalling the total capacity of each sublevel (s = 2 × 1 = 2; p = 2 × 3 = 6; d = 2 × 5 = 10; f = 2 × 7 = 14). Examination of Table 6.4, p. 144, shows us that this procedure is not infallible. Electrons occasionally show up in unexpected places (look at Cr and Cu, for example), but this need not concern us now.

Sometimes, we need to go one step further and give the *orbital diagram* of an atom, which indicates the number of electrons in each orbital and their relative spins. To derive orbital diagrams from electron configurations, we need only take account of two factors: (1) two electrons in the same orbital have opposed spins; (2) whenever possible, orbitals within the same sublevel will be half filled with electrons, all of which will have the same spin (Hund's rule). Thus, the orbital diagram for nitrogen is

1s	2s	2p
↑↓	↑↓	↑ ↑ ↑

The electron configurations of atoms correlate very simply with the positions of the corresponding elements in the periodic table. Elements in the same group of the table have the same number of electrons in the outermost principal energy level. For the A group elements, this number is equal to the group number. All the 1A elements have one "valence electron," all the 2A elements two, and so on. Since it is precisely these electrons that are involved in chemical bonding, it is hardly surprising that elements within a given group show similar chemical and physical properties.

The periodic table is a valuable device for predicting trends in properties of elements. We find, for example, that as one moves from left to right in the table, atomic radius and metallic character both decrease while ionization energy increases; as one moves down in the table, atomic radius and metallic character increase while ionization energy decreases. All of these trends, and many others, can be explained in terms of electronic structure; they are perhaps best recalled to mind by noting where elements fall in the table.

Objectives

The material covered in this chapter, unlike that of the past several chapters, is primarily non-mathematical. If you have mastered the concepts introduced here, you should be able to

1. explain how the separation between energy levels can be deduced from spectra;

2. predict which systems will behave classically and which will require quantum mechanical concepts to describe their behavior;

3. give the total electron capacity of each principal energy level, sublevel, and orbital;

4. state the rules governing the assignment of the four quantum numbers;

5. write down a reasonable set of quantum numbers for each electron in an element such as nitrogen;

6. write electronic configurations and orbital diagrams for elements such as nitrogen, sulfur, or titanium (a periodic table will be very helpful here);

7. predict the physical properties (e.g., melting point, boiling point, density) of an element, given the corresponding properties of the surrounding elements in the periodic table;

8. predict the formulas of binary or ternary compounds, given the formulas of analogous compounds formed by elements in the same group of the periodic table;

9. use the periodic table to correlate variations in atomic radius and properties that depend upon atomic size.

SELF-TEST

True or False

1. The volume of an atom is essentially that volume occupied ( )
by the electrons.

2. All the electrical charge in an atom is confined to the ( )
nucleus.

3. The atomic number is always equal to the number of ( )
electrons in a particular atom.

4. The radius of a negatively charged monatomic ion is always ( )
smaller than the radius of the parent neutral atom.

5. The energy associated with an electron in a given atom is ( )
almost fully described by specifying its value of the quantum number
n.

6. An orbital diagram is a geometrical representation of the ( )
shape of an orbital.

7. Without quantum theory, there probably would be no way of ( )
predicting the properties of element 110, yet to be discovered.

8. The energy associated with electromagnetic radiation ( )
increases in the order: x-ray, ultraviolet, visible, infrared.

Multiple Choice

9. What kind of attractive force seems to hold together the ( )
components of an individual atom?
(a) gravitational   (b) magnetic
(c) electrical   (d) the chemical bond

10. Experimental support for the arrangement of electrons in ( )
distinct energy levels is based primarily upon
(a) the law of definite (constant) proportions
(b) the law of conservation of energy
(c) x-ray spectra of atoms
(d) spectra from gas discharge

11. What is the total electron capacity of the energy level for ( )
which n = 4?
(a) 8   (b) 16   (c) 18   (d) 32

12. The possible values of the magnetic quantum number m_ℓ of a ( )
3p electron are:
(a) 0,1,2   (b) 1,2,3   (c) 1,0,–1   (d) 2,1,0,–1,–2

13. The element whose neutral, isolated atoms have three half- ( )
filled 2p orbitals is:
(a) B   (b) C   (c) N   (d) O

14. Which one of the following species has the same electron ( )
configuration as the argon atom?
(a) Ne   (b) Na^+ (c) S^- (d) Cl^-

15. The electronic configuration of the oxide ion, O^{2-}, may be ( )
represented as:
(a) $1s^2 2s^2 2p^4$ (b) $1s^2 2s^2 2p^2 3s^2 3p^2$ (c) $1s^2 2s^2 2p^6$ (d) $:O:^{2-}$

16. If a sulfur atom (atomic number 16) were to gain or lose one ( )
or more electrons to form an ion, the charge on the ion would most
likely be:
(a) +2   (b) –2   (c) –6   (d) +16

17. If ions were formed in the combination of tin ($_{50}$Sn) and ( )
another element, what would be the likely charge on the tin atoms?
(a) –2   (b) +1   (c) +3   (d) +4

18. Which one of the following species would have an odd ( )
number of electrons?
(a) N   (b) N^+ (c) NO_2^+ (d) electrons are always paired

19. Of the outermost electrons of the noble "inert" gases, it is ( )
true that:
  (a) all are in filled p orbitals
  (b) all are paired
  (c) they complete a principal energy level
  (d) all are very difficult to remove from the atom

20. Which of the following species would you expect to have the ( )
largest radius?
  (a) $_{11}Na^+$ (b) $_{10}Ne$ (c) $_9F^-$ (d) $_8O$

21. Which of the elements listed below would you expect to have ( )
the smallest atomic volume?
  (a) P   (b) Cl   (c) F   (d) all the same

22. The element with the largest value for the first ionization ( )
energy is located in the _____ portion of the modern extended
form of the periodic table.
  (a) upper right   (b) lower right
  (c) middle   (d) lower left

23. Which one of the following properties of an element would ( )
you expect *not* to be a periodic function of atomic number?
  (a) atomic volume   (b) specific heat
  (c) ionization potential   (d) boiling point

24. To estimate the density of hafnium, $_{72}Hf$, one would expect ( )
to average the densities of
  (a) La and Ta   (b) Zr and Ce   (c) Zr and U   (d) Ta and Lu

25. The atomic number of the next-to-be-discovered noble gas, ( )
below randon, Rn, would be:
  (a) 109   (b) 118   (c) 173   (d) 222

SELF-TEST ANSWERS

1. **T** (For questions 1-4, review Chapter 2.)
2. **F**
3. **F** (Consider any monatomic ion.)
4. **F** (Larger.)
5. **T**
6. **F**
7. **F** (Consider that the "periodic law," and predictions based on it, has a hundred year history.)
8. **F** (Energy and frequency decrease; wavelength increases.)
9. **c** (Nucleus-electron interactions involve opposite charges.)

10. **d**

11. **d** (Filled, it would be: $4s^2 4p^6 4d^{10} 4f^{14}$.)

12. **c** (Corresponding to the three orientations of the p orbitals: p_x, p_y, and p_z.)

13. **c** (With configuration: $1s^2 2s^2 2p^3$.)

14. **d** (Eighteen electrons.)

15. **c** (Total of 10 electrons, filling lowest available levels.)

16. **b** (High ionization potential suggests gain rather than loss of electrons; two electrons gained complete a sublevel, the 3p orbitals. See Chapter 7.)

17. **d** (The first four ionization energies would all be relatively small. The four outer-level electrons, analogous to the four in carbon, should be relatively easy to remove.)

18. **a** (Seven electrons in neutral atom = atomic number. NO_2^+ would have a total of $7 + 2(8) - 1 = 22$.)

19. **b** (Consider, with regard to choice *a*, helium. Note that you expect the first ionization energy for Rn to be rather low.)

20. **c** (Smallest nuclear charge exerts least attraction for the same number of electrons — for choices *a-c*; oxygen has fewer electrons.)

21. **c** (Outermost electrons are in a lower principle energy level.)

22. **a**

23. **b** (That of the metals is, more or less, uniformly different from that of the nonmetals. Dulong and Petit's work applied only to metallic elements.)

24. **d** (Elements 71 and 73.)

25. **b** (Element 104 would lie below $_{72}$Hf ...)

SELECTED READINGS

Drago, R. S., *Qualitative Concepts from Quantum Chemistry*, Tarrytown-on-Hudson, N. Y., Bogden and Quigley, 1971.

  A rather quantitative *treatment of atomic and molecular structure; includes exercises with answers. Goes well beyond the text in treating the quantum mechanical model of atoms and bonding.*

Hochstrasser, R. M., *Behavior of Electrons in Atoms: Structure, Spectra, and Photochemistry of Atoms*, New York, W. A. Benjamin, 1964.

  The experimental background of modern atomic structure theory is the subject of this well-written book. About the same level as the text.

Lagowski, J. J., *The Structure of Atoms*, Boston, Houghton Mifflin, 1964.

  Particularly useful are the discussions of the Bohr atom and of atomic spectra; with quotes from the pioneers.

Lewis, G. N., *Valence and the Structure of Atoms and Molecules,* New York, Dover, 1966.

> *A reprint of the 1923 classic, this presents a lucid and historical perspective on theory and experiment up to the early 1920's. It is to Lewis that we credit the idea of a bond as being a pair of electrons; his, too, the idea of the "electron dot diagrams."*

Pimentel, G. C., and R. D. Spratley, *Chemical Bonding Clarified through Quantum Mechanics,* San Francisco, Holden-Day, 1969.

> *The first two chapters are an easy reading and interesting (often amusing) introduction to the modern theory of the electronic structure of atoms.*

Seaborg, G. T., Prospects for Further Considerable Extension of the Periodic Table, *J. Chem. Ed.* (October 1969), pp. 626-634.

> *Predictions are made concerning the preparation and properties of elements not yet known, by one who has discovered his share of the elements.*

7 CHEMICAL BONDING

QUESTIONS TO GUIDE YOUR STUDY

1. Why do bonds often form between neutral atoms (as well as between ions)? Why don't all the atoms in the universe bond together in one super molecule?

2. Is there anything common to all "types" of bonds, whether ionic, covalent, metallic, etc.?

3. How do you account for the observed differences in the strengths of bonds?

4. How are the size and shape of a molecule related to the electronic structures of the component atoms? Is there a simple, reliable way of predicting the geometry?

5. Can you predict formulas of compounds on a theoretical basis; in particular, in terms of the electronic structures of the atoms involved? Can you predict how many bonds an atom may form?

6. How does bonding theory help us explain such properties of materials as color, electrical conductivity, and boiling point?

7. How would you experimentally determine bond energies and distances?

8. How would you account for the fact that an atom may bond to just one other atom, sometimes to two others, or three (e.g., C in CO, CO_2, CH_4)?

9. You have "constructed" individual atoms by filling atomic orbitals. Can you construct a molecule in an analogous manner?

10. Can you now give a more detailed description as to what occurs at the molecular level when a chemical reaction takes place?

11.

12.

YOU WILL NEED TO KNOW

Concepts

1. How to write electronic structures for representative elements — neutral atoms and also monatomic ions (say, for elements of atomic number 1 to 30, and others by analogy) — Chapter 6

2. Which means that you need to be able to use the periodic table as a correlating tool (Example: the "valence electron" structure for $_{83}$Bi is analogous to that of Sb and the other elements in Group 5A of the table: Bi — $6s^2 6p^3$; Sb — $5s^2 5p^3$) — Chapter 6

3. The shapes and relative sizes of atomic orbitals — Chapter 6

Math

1. How to calculate heats of bond formation from other heats (enthalpies) of reaction, and vice versa — Chapter 4

2. A familiarity with the geometry (symmetry and angles) of several plane and solid figures, particularly the tetrahedron, would be helpful — See a math text or the Readings of Chapter 9

CHAPTER SUMMARY—OBJECTIVES

The three types of chemical bonds that hold atoms together in elementary and compound substances are:

1) *Ionic bonds* between positive and negative ions. Ions are formed by the transfer of electrons from an element of low electronegativity (a metal) to an element of high electronegativity (a nonmetal). When two elements differ in electronegativity by more than 1.7 units, we expect the bonding between them to be predominantly ionic. We can readily deduce the charges of ions having a noble gas structure (1A, +1; 2A, +2; Al and 3B, +3; 6A, −2; 7A, −1). The transition metals commonly form cations with charges of +2 (Zn^{2+}, Cu^{2+}, Ni^{2+}, Fe^{2+}, Co^{2+}), +3 (Cr^{3+}, Fe^{3+}, Co^{3+}), or less frequently +1 (Ag^+). Many of the most common anions are polyatomic (OH^-, NO_3^-, SO_4^{2-}, CO_3^{2-}, PO_4^{3-}); the NH_4^+ ion is one of the very few cations in this category. Formulas of ionic compounds are readily deduced by imposing the condition of electroneutrality (e.g., $Zn(OH)_2$, $Cr(NO_3)_3$, Ag_2SO_4).

2) *Covalent bonds,* which consist of a pair of electrons shared between two atoms. We can expect covalent bonds to be formed when a nonmetal combines with another element which differs from it in electronegativity by less than 1.7 units. The greater the difference in electronegativity, the more

polar will be the bond. In H_2 and Cl_2 where $\Delta EN = 0$, the covalent bond is nonpolar; i.e., the electron pair is equally shared by the bonded atoms. In contrast, the bond in the HCl molecule is polar; the bonding electrons are displaced toward the more electronegative Cl atom. The displacement of electrons that characterizes a polar covalent bond strengthens and shortens it; the formation of double bonds (two pairs of electrons between bonded atoms) or triple bonds (three pairs of electrons) has the same effect.

3) *Metallic bonds,* found in elementary substances of low electronegativity (metals) and in certain alloys of these elements. Metal atoms have too few valence electrons to form covalent bonds with all of their neighbors. Instead, valence electrons are "pooled" to form a negatively charged "glue" that holds the metal cations together. This simple picture of metallic bonding, which implies high electron mobility, offers a reasonable qualitative explanation of many of the general properties of metals.

A major portion of this chapter is devoted to the structure of molecules and polyatomic ions, both of which are held together by covalent bonds. Here, we are primarily interested in two factors: the way in which the valence electrons are distributed and the angles between the covalent bonds. The electron distributions in a wide variety of polyatomic ions and molecules can be represented satisfactorily by Lewis structures; these in turn can be derived by following the simple rules listed on p. 176 of the text.

Once you have arrived at the Lewis structure for a covalently bonded species, it is relatively easy to predict the bond angles and hence the geometry of the molecule or polyatomic ion. To do this, we apply the simple principle that electron pairs surrounding an atom are oriented to be as far apart as possible. This implies, for example, that four pairs of electrons around an atom will be directed toward the corners of a regular tetrahedron. The principle is readily extended to molecules containing multiple bonds by noting that, so far as geometry is concerned, a multiple bond behaves as if it were a single electron pair.

Knowing the geometry of a molecule and the relative electronegativities of its atoms, we can decide whether it is polar or nonpolar. Molecules in which polar bonds are arranged unsymmetrically with respect to one another (e.g., H_2O, NH_3) are polar. Molecules in which all the bonds are nonpolar (e.g., H_2, Cl_2) or in which all the polar bonds "cancel" one another (e.g., CO_2, CCl_4) are nonpolar.

The concepts that we have just reviewed are fundamental to an understanding of chemical bonding and molecular structure. There are certain other less fundamental ideas with which you should be familiar. These include the terms "resonance" and "hybridization" which were grafted on to the atomic orbital approach to chemical bonding to rationalize some of its shortcomings. Finally, there is included in this chapter an introduction to a somewhat more fundamental approach to molecular structure, the molecular orbital method. Those of you who go on to take further courses in chemistry will hear a great deal more about molecular orbitals; in this chapter, we have done little more than outline the basic approach and apply it to some very simple molecules.

Objectives

After completing this chapter you should be able to:

1. predict the formulas of ionic compounds and write balanced equations for their preparation from the elements, given a periodic table (see Problem 7.4);

2. use a table of electronegativities to predict bond polarities and extent of ionic character;

3. describe and explain the factors that influence the length and strength of a covalent bond;

4. draw Lewis structures for a variety of molecules and polyatomic ions (a molecular model set, perhaps of the ball-and-stick type, will be very helpful here);

5. predict bond angles and molecular geometries, starting with the Lewis structure of a species;

6. predict whether a molecule will be polar or nonpolar, given or having derived its geometry;

7. knowing the Lewis structure of a species, predict whether it will show resonance;

8. identify the types of hybrid bonds present in a molecule, given its Lewis structure;

9. give the molecular orbital configuration for a simple diatomic species analogous to those discussed in the text;

10. interpret the properties of metals in terms of the electron sea model.

SELF-TEST

True or False

1. Chemical bonds form because one or more electrons are   (   )
attracted simultaneously by two or more nuclei.

2. Chemical bonds almost never form unless half-filled orbitals   (   )
are available.

3. A reasonable formula for a compound of aluminum and   (   )
bromine would be Al_3Br.

4. The strength of the attractive force between two bonded ( )
atoms increases, and the length of the bond increases, in the order:
single bond – double bond – triple bond.

5. The coefficient of boron, $_5B$, in the equation for the reaction ( )
of boron with oxygen, can be predicted from the electronic
structures of boron and of oxygen.

6. A binary compound consisting of an element having a low ( )
ionization potential and a second element having a high electronegativity is likely to possess covalent bonds.

7. Knowing that a Lewis structure can be written for the ( )
compound XY_2, where Y corresponds to an element in Group 7 A
of the periodic table, and that the octet rule is obeyed, you can be
reasonably sure that the neutral atom of element X has two unpaired
"valence" electrons.

8. Experimental results support the idea that a certain molecule, ( )
AB_2, is linear (that is, all three nuclei lie along a straight line). This
must mean that there are only two pairs of electrons about the
central atom.

Multiple Choice

9. The ion, Ni^{2+}, would have the electron configuration: ( )
(a) [Ar] $3d^8 4s^2$ (b) [Ar] $3d^8$
(c) [Ar] $3d^7 4s$ (d) [Ar] $3d^6 4s^2$

10. Of the elements listed below, which would you expect to ( )
have the largest attraction for bonding electrons?
(a) $_3Li$ (b) $_{13}Al$ (c) $_{26}Fe$ (d) $_{16}S$

11. The forces most suited to account for the fact that atoms ( )
often combine to form molecules are
(a) nuclear   (b) magnetic   (c) electrical   (d) polar

12. Which one of the following contains both ionic and covalent ( )
bonds?
(a) NaOH   (b) HOH   (c) C_6H_5Cl (d) SiO_2

13. For the reactions: ( )

$$C(g) + 4 H(g) \rightarrow CH_4(g) \qquad \Delta H = -395 \text{ kcal/mole}$$
$$2 C(g) + 4 H(g) \rightarrow H_2C{=}CH_2(g) \qquad \Delta H = -541 \text{ kcal/mole}$$

The enthalpy of bond formation of the C–C bond, in kcal/mole, is: ( )
(a) -99 (b) $+146$ (c) -170 (d) -146

14. Which one of the following species would have an unpaired ( )
electron?

    (a) SO_2 (b) NO_2^- (c) NO_2^+ (d) NO_2

15. Which of the following bonds would be the least polar? ( )

    (a) H—F   (b) O—F   (c) Cl—F   (d) Ca—F

16. The fact that all bonds in SO_3 are the same is accounted for ( )
by

    (a) the idea of resonance
    (b) the Lewis octet rule
    (c) electron repulsion theory
    (d) sulfur always forms three bonds

17. Indicate which one of the following does not obey the octet ( )
rule:

    (a) NO_3^- (b) O_3 (c) HCN   (d) NO_2

18. The molecular shape of the compound PH_3 is predicted as ( )
being

    (a) planar   (b) linear   (c) pyramidal   (d) something else

19. Which species is most likely to be planar? ( )

    (a) NH_4^+ (b) CO_3^{2-} (c) SO_3^{2-} (d) ClO_3^-

20. The electron pairs on Cl can be considered as being approxi- ( )
mately tetrahedrally arranged in

    (a) ClO^- (b) ClO_2^- (c) ClO_4^- (d) all of these

21. Indicate which one of the following is definitely polar: ( )

    (a) O_2 (b) CO_2 (c) BF_3 (d) C_2H_3F

22. The fact that the BeF_2 molecule is linear implies that the ( )
Be—F bonds involve

    (a) sp hybrids   (b) sp^2 hybrids
    (c) dsp^2 hybrids   (d) resonance

23. sp^3 hybridization is most likely not involved in a description ( )
of

    (a) H_3O^+ (b) CCl_4 (c) NH_4^+ (d) C_2H_2

24. Although it is known that the actual bonding sequence of ( )
atoms in the nitrous oxide molecule is N—N—O, an electronic
structure based on the sequence N—O—N (i.e., oxygen bonded to two
nitrogens):

    (a) cannot be drawn
    (b) can also be drawn, and predicts a polar molecule
    (c) can also be drawn, and involves unpaired electrons
    (d) can also be drawn, and obeys the octet rule

25. The hybridization of boron in B_2H_4 is expected to be   ( )
    (a) sp   (b) sp^2 (c) sp^3 (d) dsp^2

26. What type of hybrid orbitals are used by carbon in formal-   ( )
dehyde, CH_2O?
    (a) sp   (b) sp^2 (c) sp^3 (d) d^2sp^3

27. Metallic substances are distinguished from other classes of   ( )
chemical substances by their
    (a) high boiling points
    (b) being solids at room temperature
    (c) ability to conduct electricity in the liquid state
    (d) high conductivity in the solid state

28. The description of the electronic structure of O_2 that best   ( )
accounts for both the bond energy and the magnetic properties is
given by
    (a) :Ö=Ö:
    (b) :Ö–Ö:
    (c) a resonance hybrid of structures (a) and (b)
    (d) molecular orbital theory

29. Of the following interactions between two atoms or   ( )
molecules or ions, the strongest is most likely to be
    (a) gravitational
    (b) ionic bond
    (c) dipole-dipole forces between polar molecules
    (d) hydrogen bond

30. All chemical bonds are the result of   ( )
    (a) the magnetic interaction of electrons
    (b) the interaction of nuclei
    (c) differences in electronegativity
    (d) the interaction of electrons and nuclei

31. The electrons generally involved in bonding   ( )
    (a) are those that lie closest to the nucleus
    (b) are those for which the ionization energies are small
    (c) end up being transferred from one atom to another
    (d) occupy s atomic orbitals

32. The fact that ionization energies for "valence" electrons in   ( )
units of kcal/mole (rather than the usual eV/atom) are of the same
order of magnitude as heats of reaction for chemical changes:
    (a) is a mere coincidence
    (b) supports the idea that only the outer-level electrons are
        generally involved in molecular rearrangements
    (c) supports the idea that chemical changes involve moles of
        atoms and not simply individual atoms
    (d) has no apparent rationale

SELF-TEST ANSWERS

1. **T**

2. **F** (Look up "coordinate covalent bond". Note that hybridization often results in the unpairing of electrons. Consider: Be in BeF_2)

3. **F** ($AlBr_3$, where ions expected are Al^{3+} and Br^-.)

4. **F** (Bond length decreases.)

5. **T** (Consider atom ratio of B to O needed for octets; or, consider what ions are likely to form.)

6. **F** (Ionic: combination is of a metal and a nonmetal.)

7. **T** (Expected Lewis structure is :Ẋ:)

8. **F** (Example of reasonable structure: :B̈=A=B̈:)

9. **b** (Electrons are removed first from the highest principal energy level.)

10. **d**

11. **c**

12. **a** (Na^+ is bonded ionically to $O-H^-$, which itself is held together by a covalent bond.)

13. **d**

14. **d**

15. **b**

16. **a**

17. **d** (An odd number of electrons.)

18. **c** (With P at the center of a tetrahedron, H atoms at the base.)

19. **b** (Apply "electron pair repulsion" ideas.)

20. **d**

21. **d**

22. **a**

23. **d** (Rather, sp hybrids.)

24. **d** (For example: :N̈=O=N̈:)

25. **b**

26. **b** (Actual structure: H—C—H)
$$\overset{\displaystyle \|}{\underset{:O:}{}}$$

27. **d**

28. **d**

29. **b** (See Chapters 8 and 11 for further discussion of hydrogen bonding.)

30. **d** (See question 1.)

31. **b** (Outer-level or "valence" electrons.)

32. **b** (One can think of the formation of an ionic bond as involving ionization.)

SELECTED READINGS

Campbell, J. A., *Why Do Chemical Reactions Occur?*, Englewood Cliffs, N. J., Prentice-Hall, 1965.
> *The discussion of bonding points out that sharp lines cannot be drawn between bond types. (Example: observed bond energies vary almost continuously, gradually, from one extreme to the opposite.)*

Ferreira, R., Molecular Orbital Theory: An Introduction, *Chemistry* (June 1968), pp. 8-15.
> *A graphic introduction that should be well within reach.*

Lagowski, J. J., *The Chemical Bond*, Boston, Houghton Mifflin, 1966.
> *A survey of bonding theories, with quotations from the original works. Slights MO theory.*

Pauling, L., *The Nature of the Chemical Bond*, 3rd Ed., Ithaca, N. Y., Cornell, 1960.
> *A very extensive discussion of valence bond theory and its applications — by its creator. Written with style (almost unique in modern scientific literature). This classic is mainly for the advanced student.*

Selig, H. and others, The Chemistry of the Noble Gases, *Scientific American* (May 1964), pp. 66-77.
> *Discusses in a conversational way the discovery of the compounds of the noble gases and their bonding. Also reviews bonding theory.*

Ward, R., Would Mendeleev Have Predicted the Existence of XeF_4?, *J. Chem. Ed.* (May 1963), pp. 277-279.
> *The author's answer is Yes, and more! An interesting, compact synthesis of the ideas found in Chapters 6 and 7 of the text.*

Also see the readings of Chapter 6, by Pimentel and by Lewis.

8 PHYSICAL PROPERTIES OF MOLECULAR SUBSTANCES;NATURE OF ORGANIC COMPOUNDS

QUESTIONS TO GUIDE YOUR STUDY

1. How does modern chemical theory explain the properties of bulk samples of matter? What factors at the atomic-molecular level determine these properties?

2. What are the distinguishing properties of covalently bonded species? How are they explained? What kind of experimental observation would tell you a substance was covalently bonded?

3. What kind of experimental evidence would tell you that you were dealing with interatomic forces? intermolecular forces? (What are their relative magnitudes?) With very big or very small molecules?

4. What correlations of properties and molecular structure can be made? (For example: b.p. and shape or mass.)

5. How do you account for the existence, on the one hand, of small, discrete molecules and, on the other, of very large, extensive molecules? What factors favor the formation of one and not the other? Which elements tend to fall into which category in the formation of compounds?

6. How do you account for the formation of polymers? What special properties do polymers possess? How are they explained? (Think of some common examples of polymeric materials.)

7. Can you now more fully explain, in terms of interatomic and intermolecular forces, what you directly observe in a reaction (involving bulk samples of gases, liquids and solids)? Try, for example: Br_2 (liquid, brown) + CCl_4 (liquid, colorless) → Solution (liquid, brown).

8. How does the electronic nature of a bonded group of atoms, for example, C–O–H, determine the physical and chemical properties of the group?

9. How do you account for the vast number and variety of carbon compounds?

10. How is it possible for two different compounds to have the same molecular formula?

11.

12.

YOU WILL NEED TO KNOW

Concepts

1. How to write Lewis structures – Chapter 7
2. How to recognize, from a chemical formula, whether a species is likely to be molecular (i.e., has covalent or polar covalent bonds; a prediction based primarily on electronegativity differences) – Chapter 7
3. How to predict the geometry and polarity of molecules – Chapter 7

Math

Finally, a chapter without math!

CHAPTER SUMMARY—OBJECTIVES

In this chapter, we have considered three related topics, all of which are based on the concepts of chemical bonding and molecular structure introduced in Chapter 7.

1) *Intermolecular attractive forces,* whose magnitude determines the physical properties of molecular substances (e.g., melting point, boiling point, vapor pressure, heat of vaporization). We recognize three different types of molecular forces:

    a) *dipole forces,* which operate between polar molecules and account for the fact that these substances ordinarily have higher melting and boiling points than do nonpolar substances of comparable molecular weight.

b) the *hydrogen bond,* a particularly strong type of dipole force, which arises between molecules in which hydrogen is bonded to a small, highly electronegative atom (O, N, F). These bonds are found in a few inorganic compounds (e.g., H_2O, NH_3, HF, HCN) and a wide variety of organic compounds, including alcohols, acids, and, perhaps most important of all, proteins.

c) *dispersion forces,* which operate between all molecules, polar or nonpolar. These forces, arising from the distortion of the electron density of a molecule by its neighbor, increase in magnitude with molecular size. This explains the general observation that melting point and boiling point increase with molecular weight.

2) The structure and properties of *macromolecular substances,* where large numbers of atoms are held together by covalent bonds to form a huge molecular aggregate. Many familiar substances fall in this category. They include the allotropic forms of carbon, silicon dioxide and related compounds (quartz, sand, glass, asbestos, and others) and such familiar organic polymers as polyethylene and nylon. We have attempted to show in this chapter how the two- or three-dimensional structure of these substances determines their properties and hence their commercial uses.

3) The structure and properties of *organic compounds,* to which a major portion of this chapter is devoted. The simplest organic compounds are the hydrocarbons, which contain only carbon and hydrogen. Among hydrocarbons, we distinguish between paraffins or alkanes, in which all the bonds between carbon atoms are single (e.g., CH_4, C_2H_6, C_3H_8, $\ldots C_nH_{2n+2}$); alkenes, where there is one double bond (e.g., C_2H_4, C_3H_6, $\ldots C_nH_{2n}$); and alkynes, with one triple bond (C_2H_2, C_3H_4, $\ldots C_nH_{2n-2}$). Still another class of hydrocarbons is made up of the aromatics, which may be considered as derivatives of benzene, C_6H_6.

Many of the most familiar organic compounds contain the three elements carbon, hydrogen and oxygen. Such compounds are conveniently classified according to the functional groups present. All alcohols contain the group $-\overset{|}{\underset{|}{C}}-OH$, all organic acids the group $-\overset{}{\underset{\overset{\|}{O}}{C}}-OH$, and so on (see Table 8.4, p. 220, text). All compounds containing a given functional group show similar physical properties, within a series of alcohols, acids, etc., there is a gradation in physical properties with increasing molecular weight. In this sense, there is a resemblance to groups of elements in the periodic table.

One reason for the multiplicity of organic compounds is the phenomenon of isomerism, which is very common in organic chemistry. We ordinarily find that for a given molecular formula there are several different compounds with quite different properties. Examples include n-butane and iso-butane (molecular formula C_4H_{10}); n-propyl alcohol and isopropyl alcohol (C_3H_8O); ethyl alcohol and dimethyl ether (C_2H_6O). Noting that each of these compounds contains at least two carbon atoms bonded to each other, we see one reason ·why isomerism is much more common in organic than in inorganic chemistry. Atoms of elements other than carbon very rarely bond to one another to give structures of the type −X−X− (Can you explain why?)

Objectives

This chapter, to an even greater extent than the two that preceded it, is primarily descriptive rather than mathematical.

1. You will be expected to become familiar with the structures and properties of many of the inorganic and organic substances discussed in this chapter. You should be able to explain how the properties of such species as diamond, graphite, silica, methane, ethylene, ethyl alcohol, acetic acid, and polyethylene can be interpreted in terms of the bonds, interatomic and intermolecular forces present.

In addition, you should be able to:

2. predict the types of intermolecular forces that will operate in a variety of molecular substances;

3. predict the relative boiling points or other physical properties of a series of molecular substances, given their molecular or structural formulas;

4. classify a hydrocarbon as to type (alkane, alkene, alkyne, aromatic), given its molecular or structural formula;

5. classify an organic compound as to type (e.g., alcohol, acid), given its molecular or structural formula;

6. write a structural formula for a portion of the molecular aggregate formed when a simple monomer or a mixture of two monomers is polymerized.

SELF-TEST

True or False

1. NaOH(s) is a poor electrical conductor because it is made up ( )
of molecules.

2. The physical properties of molecular substances are directly ( )
related to the strengths of the covalent bonds holding the molecule
together.

3. Boiling points of molecular substances usually increase with ( )
molecular weight.

4. Ethyl alcohol, C_2H_5OH, would be expected to show ( )
hydrogen bonding.

5. The molecular formula C_5H_{10} represents an alkane. ( )

6. The saturated hydrocarbon C_4H_{10} shows *cis-trans* isomerism. ( )

7. The double bonds in benzene and other aromatic hydro- ( )
carbons behave chemically in the same way as those in ethylene.

8. The compound $H_3C-C-O-C_4H_9$ is an ester. ( )
$\overset{\|}{O}$

9. In polyethylene, $(CH_2)_n$, hydrogen accounts for 2/3 of the ( )
mass.

10. An increase in pressure increases the stability of diamond ( )
relative to graphite.

Multiple Choice

11. Which one of the following solid substances consists of small, ( )
discrete molecules?
    a) graphite   b) Dry Ice   c) iron   d) NaCl

12. Of the following interactions at the atomic-molecular level, ( )
the strongest is most likely to be
    a) dispersion   b) dipole-dipole
    c) hydrogen bond   d) covalent bond

13. Which one of the following species could be boiled without ( )
breaking hydrogen bonds?
    a) CH_4 b) NH_3 c) H_2O d) none of these

14. In which pair of compounds must the same type of attractive ( )
force be overcome upon melting?
    a) LiCl and ICl   b) SiO_2 and CO_2
    c) CCl_4 and O_2 d) K and C

15. Which of the following would have a boiling point lower than ( )
$SiCl_4$?
    a) $GeCl_4$ b) $SiBr_4$ c) CCl_4 d) LiCl

16. Which of the following hydrocarbons is most likely to be a ( )
liquid at 25°C and 1 atm? (The rest are gases).
    a) C_2H_4 b) C_2H_6 c) C_4H_{10} d) C_6H_{14}

17. Compounds of carbon are so numerous and varied because ( )
   a) carbon is the most abundant element
   b) carbon atoms readily bond to each other
   c) there are more people involved in research in organic than in inorganic chemistry
   d) every compound known contains one or another isotope of carbon

18. Which of the following compounds is an alkene? ( )
   a) $H_3C-CH_2-CH_3$ b) $H_2C=CH-CH_3$
   c) $HC\equiv C-CH_3$ d) $C_6H_5CH_3$

19. How many different compounds have the molecular formula ( )
C_5H_{12}?
   a) 1   b) 2   c) 3   d) some other number

20. Which of the following compounds is a ketone? ( )
   a) $H_3C-\underset{\underset{O}{\|}}{C}-H$ b) $H_3C-\underset{\underset{O}{\|}}{C}-OH$

   c) $H_3C-\underset{\underset{O}{\|}}{C}-CH_3$ d) $H_3C-\underset{\underset{O}{\|}}{C}-OCH_3$

21. The compounds CH_3OH and $H_3C-COOH$ would react ( )
together to produce
   a) an alcohol   b) a ketone   c) an ester   d) a fat

22. For which of the following polymers are two different ( )
monomers required?
   a) polyethylene b) nylon c) polystyrene d) none of these

23. To convert graphite to diamond, one would use ( )
   a) high temperatures at atmospheric pressure
   b) high pressures at an elevated temperature
   c) oxygen under pressure
   d) the philosopher's stone

24. The various types of glass you use in the laboratory are ( )
unlikely to contain appreciable amounts of:
   a) oxygen   b) boron   c) sodium   d) hydrogen

25. Many of the properties of graphite resemble those of mica ( )
because
   a) both contain a group 4A element
   b) both have two-dimensional layer structures
   c) both are thermodynamically stable at 25°C and 1 atm
   d) they occur together in nature

SELF-TEST ANSWERS

1. F   (Ions are fixed in position.)
2. F   (Intermolecular forces are determining factors.)
3. T   (With some exceptions, notably water.)
4. T   (Compare H–O–H and –C–O–H.)
5. F
6. F   (Double bond required.)
7. F   (Contrast substitution vs addition.)
8. T
9. F   (2/3 of atoms, 1/7 of mass.)
10. T
11. b
12. d
13. a
14. c   (Dispersion.)
15. c
16. d
17. b   (c may be true but irrelevant.)
18. b
19. c   (C–C–C–C–C; C–C–C–C with C branch; C–C–C with C branches)
20. c
21. c
22. b
23. b   (Or d, if available.)
24. d
25. b

SELECTED READINGS

Amoore, J. E., and others, The Stereochemical Theory of Odor, *Scientific American* (February 1964), pp. 42-49.
    A summary of early work on a "lock and key" model based on the geometry of molecules, with several successful tests of the theory described.

Barrow, G. M., *The Structure of Molecules: An Introduction to Molecular Spectroscopy,* New York, W. A. Benjamin, 1963.
    A detailed examination of molecular energies and how they are experimentally determined and what they tell us about molecular structure.

Breslow, R., The Nature of Aromatic Molecules, *Scientific American* (August 1972), pp. 32-40.
    An elementary but interesting discussion of the aromatic class of organic compounds (and how their molecular orbitals are filled with "magic numbers" of electrons).

Brey, W. S. Jr., *Physical Methods for Determining Molecular Geometry*, New York, Reinhold, 1965.

Discusses other tools as well as those described by Barrow, including diffraction techniques (see Chapter 9, text).

Brown, J. F. Jr., Inclusion Compounds, *Scientific American* (July 1962), pp. 82-92.

A dated but interesting discussion (there are several references to the "inert" gases) of another "lock and key" phenomenon, highly dependent on geometry.

Clapp, L. B., *The Chemistry of the OH Group*, Englewood Cliffs, N. J., Prentice Hall, 1967.

An introduction to the properties of organic compounds, with an emphasis on correlation with structure.

Herz, W., *The Shape of Carbon Compounds: An Introduction to Organic Chemistry*, New York, W. A. Benjamin, 1963.

With an emphasis on structure, a broad introduction to organic chemistry.

Materials, San Francisco, W. H. Freeman, 1967.

The September 1967 issue of Scientific American. *This is a valuable introduction to the new science of materials (polymers, glasses, ceramics, etc.). This will be useful reading also for Chapter 9.*

Wagner, J. J., Nuclear Magnetic Resonance Spectroscopy — An Outline, *Chemistry* (March 1970), pp. 13-15.

A qualitative introduction to one of the newer tools for determining molecular structure.

9 LIQUIDS AND SOLIDS; CHANGES IN STATE

QUESTIONS TO GUIDE YOUR STUDY

1. What properties do you generally associate with solids; liquids; gases? Can you qualitatively account for the differences?

2. What kinds of substances normally exist (i.e., at 1 atm, room temperature) as solids; liquids; gases?

3. Under what conditions may a given substance exist as a solid; liquid; gas?

4. Are there simple laws, as there are for gases, relating volume to temperature and pressure for a liquid or a solid?

5. How would you describe the process of evaporation; freezing? (First, in terms of what you observe; then in terms of atomic-molecular behavior.) What energy effects accompany these processes?

6. What kind of experiment would you do to show the existence of a vapor above a liquid? Or to show how the pressure of this vapor changes with temperature?

7. How would you recognize boiling? What would you measure?

8. How do we know the arrangements of molecules in solids and liquids? How do you account for the regularities of shape and size of crystalline solids? (Can you name some crystalline solids?)

9. How do you account for the fact that some solids are converted directly to gases, without first melting? (Examples: snow in the depths of winter; "dry ice".)

10. How are the properties of solids and liquids important to the activities of an architect; a ham radio operator; a man-eating shark?

11.

12.

77

YOU WILL NEED TO KNOW

Concepts

1. How many "valence electrons" there are in any given atom (a prediction that can be made from position in the periodic table) — Chapter 6
2. The ideas of kinetic theory about the nature of a gas (this chapter extends kinetic theory to liquids and solids) — Chapter 5

Math

1. How to use the ideal gas law, $pV = nRT$ (e.g., to calculate p from V, n, T) — Chapter 5
2. How to find a logarithm or an antilogarithm (with either a log table or a slide rule) — See the text (Appendix 4), a math text, or the math preparation manual suggested in the Preface
3. How to plot the equation of a straight line; how to read such a graph (e.g., $\log p = \dfrac{-\Delta H}{2.3\,RT} + B$ has the form $y = mx + b$) — See above (2)
4. The geometry of the cube: how to relate the dimensions — edge, body diagonal, face diagonal (For a cube with an edge of length α: face diagonal $= \sqrt{2}\,\alpha$, body diagonal $= \sqrt{3}\,\alpha$.)

CHAPTER SUMMARY—OBJECTIVES

The principal distinction between the gaseous state of matter, discussed in Chapter 5, and the condensed states (liquid and solid) considered in this chapter is the distance of separation between molecules. In the gas state, at ordinary temperatures and pressures, the molecules themselves account for a negligible fraction of the total volume. In the condensed states, the molecules ordinarily occupy from 50 to 70 per cent of the total volume. This structural difference explains why liquids and solids have much greater densities than gases, why they are less compressible, and expand less on heating.

Since molecules in the condensed states are closer together, short-range intermolecular forces exert a much greater influence on physical behavior. The strength of these forces depends upon the type of molecule present. This is why we cannot write a simple equation of state, analogous to the ideal gas law, to describe the physical behavior of all liquids or all solids. Each liquid and solid has a characteristic density, compressibility, and expansibility.

The major distinction between the liquid and solid states is one of molecular mobility. The particles in a liquid are relatively free to move with respect to one another, while the particles in a solid are restricted to small vibrations about a point in the crystal lattice. This makes the particle structure of solids much easier to study experimentally than that of liquids. The technique of x-ray diffraction can be applied to find the basic structural unit of the crystal, the unit cell; x-ray diffraction patterns for liquids are diffuse and difficult to interpret.

The particles in a solid tend to pack as closely as possible. Metals, in which all the atoms are identical, tend to crystallize in one of the two closest-packed arrangements (cubic or hexagonal). Another common type of packing is body-centered cubic, where the fraction of empty space is only a little greater. In ionic crystals, where the two kinds of ions differ in size, there is a greater variety of packing patterns. Frequently, the larger anions form a close-packed array, slightly expanded if necessary to accommodate the smaller cations fitting into "holes" in the anionic framework.

Much of this chapter was spent in discussing transitions from one state of matter to another. The nature of the equilibria involved can perhaps best be summarized by a phase diagram such as that shown on p. 251 of the text. At any given temperature, a liquid or solid has a characteristic vapor pressure (curves AB and AC, Figure 9.12). For all liquids and solids, vapor pressure rises exponentially with temperature; the general equation is:

$$\log_{10} p = \frac{-\Delta H}{(2.30)\,(1.99)\,T} + constant$$

where p is the vapor pressure, T the absolute temperature, and ΔH the enthalpy change for the transition (heat of vaporization or sublimation, in cal.).

Two features of liquid-vapor equilibria which are not apparent from Figure 9.12 are the phenomena of boiling and critical behavior. A liquid boils (i.e., vapor bubbles form within the body of the liquid) when its vapor pressure becomes equal to the applied pressure; the normal boiling point is the temperature at which the vapor pressure becomes one atmosphere. A liquid can be heated in a closed container to temperatures far above its boiling point. Eventually, however, the kinetic energy of the molecules becomes too great for them to remain in the liquid; at this so-called "critical" temperature, the liquid suddenly and spontaneously vaporizes.

Referring to the phase diagram on p. 251, we see that there is one particular temperature and pressure (the triple point) at which all three states of a substance are in equilibrium with each other. For most substances, the triple point pressure is relatively low. For a few substances, including CO_2, the triple point pressure exceeds one atmosphere. Such substances, when heated in an open container, pass directly from the solid to the vapor state (sublimate).

The line labeled AD in Figure 9.12 describes the equilibrium between the solid and liquid states. To the extent that this line deviates from the vertical, the melting point of the substance changes with pressure. For most substances, where the solid phase is the more dense, an increase in pressure favors the formation of solid and the melting point rises. For water, where the liquid is the more dense phase, the reverse is true. The effect is always small; pressures in the range of 50 to 100 atmospheres are required to change the melting point by as much as 1°C.

Objectives

After completing this chapter, you should be able to:

1. explain, from a molecular viewpoint, the differences in physical behavior between gases, liquids, and solids;

2. explain, on a molecular basis, what happens when a substance changes from one state to another;

3. explain what is meant by boiling point; critical temperature; critical pressure; triple point; sublimation;

4. have sufficient understanding of the concept of vapor pressure to answer questions such as Problems 9.7 and 9.8;

5. use Equation 9.3 to determine graphically the heat of vaporization of a liquid, knowing its vapor pressure at a series of temperatures;

6. use Equation 9.4 to calculate the variation of vapor pressure with temperature, given the heat of vaporization;

7. be sufficiently familiar with the geometry of cubic cells to make calculations of the type called for in Problems 9.16 and 9.17;

8. draw a phase diagram for a pure substance, given appropriate data, and identify each area, line, and point.

SELF-TEST

True or False

1. At 50°C, the vapor pressure of liquid A is found to be 50 mm ( )
Hg; that of liquid B, 125 mm Hg. One can be reasonably sure that liquid A has the higher normal boiling point.

2. For these same two liquids, one can state that the liquid with ( )
the higher surface tension at 50°C is probably A.

3. For any pure substance, the melting point is always a little ( )
above the freezing point. For instance, water will melt at a temperature above 0°C and freeze at a temperature a little below 0°C.

4. Not all solids are crystalline. ( )

5. Whenever bubbles form within a sample of liquid, the liquid ( )
is said to be boiling.

6. Generally, when a solid melts to form a liquid, the density (T) H_2O is
decreases. exception

7. For all solids and liquids, vapor pressure increases linearly (F) exponentially
with temperature.

8. Below the triple point temperature, only solid can exist. ( )

9. Pure water cannot exist as a liquid below 0°C. F( )
supercooling

Multiple Choice

10. When one mole of liquid water is placed in a 100 ml flask at ( )
25°C, it eventually establishes a constant pressure of 24 mm Hg. If a
200 ml flask were used instead, the final pressure would have been:
(a) 12 mm Hg (b) 48 mm Hg (c) 24 mm Hg (d) 760 mm Hg

11. Which substance should have the highest vapor pressure at ( )
any given temperature?
(a) C_2H_6 (b) C_3H_8 (c) C_4H_{10} (d) C_5H_{12}

12. The vapor pressure of CCl_4 at 77°C is 760 mm Hg. The heat ( )
of vaporization is approximately
(a) 1617 cal   (b) 3790 cal   (c) 7350 cal   (d) 9340 cal

13. The boiling point of any liquid is: ( )
(a) 100°C
(b) the temperature at which as many molecules leave the
liquid as return to it
(c) the temperature at which the vapor pressure is equal to
the external pressure
(d) the temperature at which no molecules can return to the
bulk of the liquid

14. The triple point is ( )
(a) the temperature above which the liquid phase cannot
exist
(b) usually found at a temperature very close to the normal
boiling point
(c) the temperature at which the vapor pressure of a liquid is
three times the value at 25°C
(d) the value of temperature and pressure at which solid,
liquid, and gas may exist in equilibrium

15. A certain substance, X, has a triple point temperature of ( )
20°C at a pressure of 2.0 atmospheres. Which one of the following
statements *cannot* be true?
    (a) X can exist as a liquid above 20°C
    (b) X can exist as a solid above 20°C
    (c) liquid X is stable at 25°C and at 1 atm
    (d) both liquid X and solid X have the same vapor pressure at
        20°C

16. A certain solid sublimes at 25°C and 1 atmosphere pressure. ( )
This means
    (a) the solid is more dense than the liquid
    (b) the pressure at the triple point is greater than 1
        atmosphere
    (c) the solid is less dense than the liquid
    (d) the pressure at the triple point is less than 1 atmosphere

17. The melting point of benzene at 1 atmosphere is 5.5°C. The ( )
density of liquid benzene is 0.90 g/ml; that of the solid is 1.0 g/ml.
At an applied pressure of 10 atm, the melting point of benzene
    (a) is equal to 5.5°C
    (b) is slightly greater than 5.5°C
    (c) is slightly less than 5.5°C
    (d) cannot be estimated from the information given

18. Dry ice, solid CO_2, is frequently used as a refrigerant; it ( )
undergoes sublimation at or near atmospheric pressure. This direct
conversion of solid CO_2 to gaseous CO_2 involves an absorption of
energy by the CO_2. This energy is used up, at least in part, in over-
coming
    (a) covalent bonds   (b) polar covalent bonds
    (c) gravitational attractions   (d) van der Waals forces

19. When a solid melts to a liquid ( )
    (a) greater attractive forces appear
    (b) the molecules become more randomly oriented
    (c) the molecules become less randomly oriented
    (d) the container warms up

20. The number of closest neighbors in a body-centered cubic ( )
lattice is
    (a) 4   (b) 6   (c) 8   (d) 12

21. A certain metal crystallizes in a face-centered cubic unit cell.   (   )
The relationship between the atomic radius (r) of the metal and the
length (l) of one edge of the cube is:

(a) $2r = l; r = l/2$

(b) $2r = (2)^{1/2} l; r = (2)^{1/2} l/2$

(c) $4r = (2)^{1/2} l; r = (2)^{1/2} l/4$

(d) $4r = (3)^{1/2} l; r = (3)^{1/2} l/4$

22. A chemistry handbook normally gives the pressure at which   (   )
the boiling point of a pure substance is measured but generally does
not indicate the pressure for the melting point. Why?

(a) both melting point and boiling point are always measured
at the same pressure

(b) the melting point is usually nearly independent of
pressure

(c) solids are more often impure and therefore have variable
melting points

(d) all melting points are measured at one atmosphere

23. A sample of gaseous water at a pressure of 17.5 mm Hg is in   (   )
equilibrium with liquid water at 20°C. On reducing the volume
available to the system, you observe the pressure remains at 17.5 mm
Hg. The observation is accounted for by assuming:

(a) gaseous water doesn't behave ideally under these
conditions

(b) some vapor escapes from the container

(c) your pressure measurement was incorrect; it should have
been higher

(d) some vapor condenses to form liquid

SELF-TEST ANSWERS

1. **T** (At 50°C, B is closer to the boiling point than A since its vapor pressure is higher.)

2. **T** (The lower v.p. would result from stronger intermolecular attractions; so would a higher surface tension.)

3. **F** (They are one and the same equilibrium temperature — at which both solid and liquid can exist.)

4. **T** (Many solids, such as glass and rubber, are *amorphous,* lacking the regularity of atomic placement found in crystals.)

5. **F** (Bubbles often indicate dissolved gases coming out of solution.)

6. **T** (Water is an exception.)

7. **F** (Exponentially.)

8. **F** (What about the pressure?)

9. **F** (Supercooling! Also consider higher pressures.)

10. c (Vapor pressure depends only on temperature. Could all the water have evaporated?)

11. a

12. c ($\Delta H/T = 21$, approx.)

13. c (If you chose b, consider boiling water in an open container.)

14. d

15. c (Construct a phase diagram.)

16. b

17. b

18. d (Chapter 8.)

19. b

20. c

21. c

22. b (The solid-liquid line in a phase diagram is almost vertical.)

23. d

SELECTED READINGS

Bernal, J. D., The Structure of Liquids, *Scientific American* (August 1960), pp. 124-128.
Much of the work on liquid structure has been done by the author. He makes comparisons to other states and presents a model for a state which we commonly, carelessly, think of as being without structure.

Bragg, L., X-Ray Crystallography, *Scientific American* (July 1968), pp. 58-70.
A detailed survey of the techniques and successes of x-ray diffraction, written by one who has made history in the field. About the same level as the text.

Fullman, R. L., The Growth of Crystals, *Scientific American* (March 1955), pp. 74-80.
An interesting, well-written introduction to the nature of crystal growth: the "screw-dislocation" theory.

Greene, C. H., Glass, *Scientific American* (January 1961), pp. 92-105.
Undercooled liquid or inorganic polymer? The article describes the structure of this noncrystalline solid and the processes by which it is made.

Moore, W. J., Seven Solid States: An Introduction to the Chemistry and Physics of Solids, New York, W. A. Benjamin, 1967.
Often advanced, this is a discussion of seven exemplary solids, including salt, gold, ruby and steel. Much material assembled here.

Sanderson, R. T., The Nature of 'Ionic' Solids, *J. Chem. Ed.* (September 1967), pp. 516-523.
A different approach to the solids we have labeled as bonded ionically — that emphasizes the connection with covalent bonding. At least read the summary.

Turnbull, D., The Undercooling of Liquids, *Scientific American* (January 1965), pp. 38-46.
Supercooling is a general phenomenon. This article discusses "nucleation" and the nature of the freezing process, as well as reviews the structure of liquids.

Wells, A. F., The Third Dimension in Chemistry, New York, Oxford, 1956.
This book will let you find some use for that geometry you learned long ago. Also of interest might be the huge volume Structural Inorganic Chemistry, same author, same publisher, authoritative.

10   SOLUTIONS

QUESTIONS TO GUIDE YOUR STUDY

1. What kinds of processes occur within solids; within liquids? For example, what happens at the molecular level when two liquids are mixed; or when a solid dissolves in a liquid?

2. How do the properties of a substance (such as f.p., b.p.) change when a second substance is dissolved in it? Are there simple laws relating solution properties?

3. Can you think of some naturally occurring solutions? (Recall that we generally do not encounter pure substances outside the lab.)

4. How do you quantitatively describe the composition of a solution?

5. How do the properties of a solution depend on the relative amounts of its components?

6. What properties of substances determine the extent to which one will dissolve in the other? What generalizations can be made?

7. How do changes in conditions, such as temperature and pressure, affect solubility?

8. How would you experimentally show that a given solution is saturated; supersaturated? How would you prepare such a solution?

9. Can you describe the structure, the bonding, of solutions? Can you account for any energy effects accompanying the formation of a solution?

10. To what uses are solutions put in the laboratory?

11.

12.

YOU WILL NEED TO KNOW

Concepts

1. How to predict the geometry and polarity of molecules; how to predict bond type, in general — Chapter 7

Math

1. How to express amounts of substances in the various common units: grams, moles, number of particles — Chapter 3

CHAPTER SUMMARY—OBJECTIVES

We saw in Chapter 9 that the magnitude of a substance's intermolecular forces determines whether it exists as a gas, liquid, or solid at normal temperatures and pressures (e.g., 25°C, 1 atm). Again, we might expect two substances with intermolecular forces of about the same magnitude to be infinitely soluble in each other. Putting these two generalizations together, we expect complete miscibility only if the two substances involved are in the same physical state. Experiment confirms this reasoning; all gases are infinitely soluble in one another; complete miscibility is the rule rather than the exception among liquids of similar polarity. In contrast, solids and gases always show limited solubility in liquid solvents. Carrying this reasoning one step further, we deduce that the closer a solid or gas is to the liquid state (i.e., the lower the melting point of the solid or the higher the normal boiling point of the gas) the greater will be its solubility. At the opposite extreme, solutes whose intermolecular forces differ vastly from those of the solvent will be virtually insoluble in it (e.g., the permanent gases and hydrocarbons in water).

It is, of course, possible to change the solubility of a solute by changing the external conditions of temperature and pressure. If the solution process is endothermic, an increase in temperature promotes solubility. This is almost always the case with solid-liquid systems and usually true when both components are liquids. With a gas, the solution process may be either endothermic or exothermic; the solubility of the permanent gases in organic solvents normally increases with temperature while the same gases become less soluble in water when the temperature is raised. Solubility is little affected by pressure except for gas-liquid systems where it is directly proportional to the partial pressure of the gas over the solution.

Various methods are used to express the concentrations of solutions, i.e., the relative amounts of solute and solvent. Frequently, the weight fractions are given by quoting weight per cents of the two components or, for very dilute solutions, "parts per million" or even "parts per billion" of solute (1 ppm = 1 g solute per 10^6 g solvent). In the laboratory, reagent concentrations are most frequently expressed in terms of molarity (no. of moles per liter of solution). Two other concentration units, mole fraction and molality, are particularly useful for expressing the colligative properties of solutions. The terminology here is unfortunate; students often confuse molality (m) with molarity (M), or even with morality, which is something else altogether.

Almost without exception, the addition of a small amount of solute lowers the freezing point of a solvent; if the solute is nonvolatile, the boiling point will be raised as well. Both of these effects can be related to the lowering of solvent vapor pressure that always accompanies the formation of a solution. One can also explain the phenomenon of osmosis in terms of vapor pressure lowering; water or other solvent moves through a semipermeable membrane from a region of high vapor pressure (pure solvent) to one of low vapor pressure (solution). Osmosis can be prevented by exerting sufficient force on the solution side of the membrane; the pressure required to do this is referred to as the osmotic pressure.

Each of the effects just described (vapor pressure lowering, freezing point depression, boiling point elevation, osmotic pressure) is a colligative property; its magnitude depends primarily upon the concentration of solute particles rather than their type. The appropriate equations appear in the text (Equations 10.8, 10.9, 10.10, 10.11). Notice that these equations incorporate a constant of proportionality which in every case but one is characteristic of the particular solvent. The exception is the expression for osmotic pressure, where the gas law constant R appears. (This constant keeps popping up in the most unexpected places!) These equations can be used in a very practical way to calculate vapor pressures, freezing points, boiling points, and osmotic pressures of solutions. In addition they serve as the basis for the experimental determination of the molecular weight of a solute.

Objectives

If you have mastered the material in this chapter, you should be able to:

1. calculate the weight %, mole fractions, and molality of solutions, given the masses in grams of the components;

2. use the defining equation for molarity to calculate any of three quantities (M, no. moles, volume in lit), given the other two, or solve dilution problems such as that of Problem 10.6;

3. predict relative solubilities of gases, liquids and solids in liquid solvents;

4. predict the effect of a change in temperature upon solubility in gas-liquid, liquid-liquid, and solid-liquid systems;

5. use Henry's Law to calculate the dependence of gas solubility on the partial pressure of the gas;

6. use Equations 10.8, 10.9, 10.10, 10.11, to calculate the colligative properties of solutions of known concentration;

7. use these equations to obtain the molecular weight of a solute from appropriate experimental data.

SELF-TEST

True or False

1. In a solution containing equal numbers of grams of benzene ( )
(MW = 78) and toluene (MW = 92), the mole fractions are each 0.50.

2. A supersaturated solution of air in water could be prepared ( )
by bubbling air through water at 60°C and cooling to 25°C.

3. The mole fractions of all the components of a solution add to ( )
unity.

4. If water saturated with nitrogen at 1 atm is exposed to air, ( )
N_2 will come out of solution.

5. DDT should be more soluble in alcohol than in carbon tetra- ( )
chloride.

6. The solubility of a solid in a liquid is ordinarily directly ( )
proportional to the absolute temperature.

7. Raoult's Law often applies to the solvent in a solution but ( )
not to the solute.

8. The boiling point of a 1 m solution of a nonelectrolyte will ( )
be 100.52°C, provided the solution is ideal.

9. Freezing point lowering, boiling point elevation, and osmotic ( )
pressure can all be explained in terms of vapor pressure lowering.

10. The freezing point of a 1 m solution of K_2SO_4 would be ( )
about the same as that of a 1 m solution of urea.

Multiple Choice

11. When one mole of KCl is dissolved in a kilogram of water, the ( )
concentration of Cl^- is:
    a) 0.5 molal   b) 0.5 molar   c) 1.0 molal   d) 1.0 molar

12. A student wishes to prepare 100 cc of 0.50 M NaCl from ( )
2.00 M NaCl. What volume of the more concentrated solution should
he start with?

    a) 25 cc   b) 50 cc   c) 100 cc   d) 400 cc

13. The solubility of a certain salt in water is 22 g/liter at 25°C ( )
and 60 g/liter at 80°C. A student prepares 500 cc of saturated
solution at 80°C, cools to 25°C, and adds a tiny crystal of salt. How
many grams of salt come out of solution?

    a) 11   b) 19   c) 22   d) 8

14. The number of grams of Na_2SO_4 (FW = 142) required to ( )
prepare 2.00 liters of 1.50 M solution is:

    a) 3   b) 213   c) 142   d) 426

15. To obtain 12.0 g of K_2CrO_4 from a solution labeled "5.0% ( )
K_2CrO_4 by weight", how many grams of solution should you weigh
out?

    a) 2.4   b) 5.0   c) 12   d) 240

16. Which of the following has the least effect on the solubility ( )
of a solid in a liquid solvent?

    a) nature of solute   b) nature of solvent
    c) temperature   d) pressure

17. The solubility of CO_2 in water at 20°C is 3.4 g/liter at 1 atm. ( )
If the CO_2 pressure is raised to 3 atm, the solubility, in g/liter, is
expected to be

    a) 3.4/3   b) 3   c) 3.4   d) 3.4 X 3

18. For which of the following pairs would you expect solubility ( )
to be the greatest?

    a) O_2-water b) sugar-water c) sugar-benzene d) O_2-N_2

19. A student wants to remove naphthalene, an aromatic hydro- ( )
carbon, from his lab coat. What solvent would you recommend?

    a) water b) ethyl alcohol c) benzene d) sulfuric acid

20. When one mole of a nonvolatile nonelectrolyte is dissolved in ( )
two moles of a solvent, the vapor pressure of the solution, relative to
that of the pure solvent is:

    a) 1/3   b) 1/2   c) 2/3   d) cannot tell

21. Which one of the following is not a colligative property? ( )

    a) freezing point   b) molarity
    c) osmotic pressure   d) color

22. The osmotic pressure of a 1 M solution of sugar at room ( )
temperature would be closest to:
    a) 0.25 atm   b) 1 atm   c) 2.5 atm   d) 25 atm

23. To determine the molecular weight of a polymer (appr. MW = ( )
10,000) one would probably measure
    a) osmotic pressure of a solution
    b) boiling point of a solution
    c) density of the solid
    d) vapor pressure of a solution

24. A small amount of a certain solute melting at 800°C is added ( )
to water. The solution will be expected to freeze
    a) above room temperature   b) slightly above 0°C
    c) at 0°C   d) slightly below 0°C

25. In a 0.1 M solution of NaCl in water, which one of the ( )
following will be closest to 0.1?
    a) mole fraction NaCl   b) mole fraction water
    c) wt. % NaCl   d) molality

SELF-TEST ANSWERS

1. F (Larger no. moles benzene.)
2. F (Solubility increases as temperature drops.)
3. T
4. T (p_{N_2} reduced.)
5. F
6. F (Relationship not linear.)
7. T
8. F (Depends on pressure, solute volatility.)
9. T
10. F (More solute particles with K_2SO_4.)
11. c
12. a
13. b (I.e., 30—11.)
14. d
15. d (I.e., 12.0/0.050.)
16. d
17. d
18. d (Same physical state.)
19. c
20. c
21. d

22. d
23. a   (Most sensitive.)
24. d
25. d

SELECTED READINGS

Dye, J. L., The Solvated Electron, *Scientific American* (February 1967), pp. 76-83.
   For many years a lab curiosity, it now appears that the solvated electron may be an intermediate reactive species in many reactions.
Mysels, K. J., The Mechanism of Vapor-pressure Lowering, *J. Chem. Ed.* (April 1955), p. 179.
   Points out errors in the argument usually given (but doesn't offer anything in its place).

   For working problems pertaining to solutions (concentrations, reaction stoichiometry, colligative properties, etc.) see problem manuals listed in the Preface.

11 WATER, PURE AND OTHERWISE

QUESTIONS TO GUIDE YOUR STUDY

1. Where and in what physical state do you find water, near and not so near the earth?

2. Is pure water, like many other pure substances, found only in the laboratory?

3. What's so special about water? For example: How does the heat required for changes in state compare to that of other substances? How does the melting point compare to that for a substance of similar molecular weight and geometry? How do its solvent properties compare to those of other liquids?

4. What geometry and bonding do you predict for the water molecule? How does this structure help explain water's unique properties?

5. How do water solutions compare to nonaqueous solutions? Do the same laws of colligative properties and the same principles of solubility apply?

6. What solutes are found in naturally occurring water solutions? What are their sources?

7. How can you measure the concentration of a solute in water solution?

8. How can you prepare very pure water? How is drinking water normally prepared? (What processes are used in a municipal water plant?)

9. Do we have, as we do for mixtures of gases, a workable model for simple salt solutions?

10. Can you justify an extensive study of water solutions? (The remaining chapters in the text are devoted almost entirely to reactions which occur in water solution.)

11.

12.

YOU WILL NEED TO KNOW

Concepts

1. The kinds and relative strengths of forces that exist between molecules (e.g., hydrogen bonding) — Chapter 8
2. How to write and interpret chemical equations — Chapter 3

Math

1. How to work problems involving concentration units (e.g., calculating mass of solute from volume and concentration of solution) — Chapter 10
2. How to perform stoichiometric calculations (those based on chemical equations) — Chapter 3

CHAPTER SUMMARY—OBJECTIVES

The unique properties of water (e.g., high boiling point and heat of vaporization, volumetric behavior near 0°C) are related to the ability of the H_2O molecule to form three-dimensional networks held together by hydrogen bonds. Such networks are known to exist in ice, accounting for its low density relative to liquid water. They are believed to persist, probably in modified form, in liquid water near the melting point. As the temperature is raised, there is a shift toward a more closely packed arrangement typical of normal liquids. Qualitatively at least, we can explain the unusual behavior of water near 0°C (minimum volume at 4°C) in terms of an equilibrium between "flickering clusters" of hydrogen bonded water molecules and a more closely packed structure.

Despite a great deal of research in the area, the effect of electrolytes on the water structure is still a matter of controversy. At low concentrations, below 0.01 m, the freezing point lowering of a solution of a 1:1 electrolyte such as NaCl can be expressed by the equation:

$$\Delta T_f = i\,(1.86°C)\,m; \quad i = 2 - 0.78\,m^{1/2}$$

If the solute ions (e.g., Na^+, Cl^-) behaved as completely independent particles, i would be equal to 2. We see from this equation that as we go to very, very dilute solutions (m → 0), where the ions are extremely far apart, i → 2. The correction term in the expression for i takes account of interactions between oppositely charged ions, the "ion-atmosphere" effect of Debye. Equations analogous to this can be written for other types of electrolytes; we always find that as m → 0, i approaches the number of moles of ions per mole of electrolyte. All of these equations break down at embarrassingly low concentrations; nevertheless, they can be applied to freezing point data in very dilute solutions to obtain i. In this way, it is possible to deduce the way in which a species ionizes in water: we can demonstrate for example that H_3PO_4 forms H^+ and $H_2PO_4^-$ ions (i = 2), while H_2SO_4 gives 2 H^+ + SO_4^{2-} (i = 3).

Relatively small amounts of suspended or dissolved species in water can make it unsafe for drinking, unappealing for recreational purposes, or harmful in other ways to the environment. One measure of the extent of pollution of water by organic material is its B.O.D. (biochemical oxygen demand), which tells us the concentration of oxidizable organics in water. Vapor phase chromatography (Chapter 1) can be used to determine pesticides which concentrate in the food chain to the point where they have become a menace to wildlife and perhaps to human beings as well. Many inorganic contaminants can have undesirable effects. An example is the phosphate ion, present in many detergents, which can contribute to the eutrophication of lakes. Compounds of certain of the heavy metals, notably mercury and lead, are known to be toxic even at very low concentrations.

Among the processes commonly used in municipal water purification are sedimentation, coagulation, filtration, and disinfection. These are designed primarily to clarify water and destroy disease-carrying organisms. Waters containing relatively high concentrations of Ca^{2+}, Mg^{2+} or Fe^{3+} can be softened by passing through columns containing natural or synthetic zeolites, which replace the objectionable cations by Na^+ ions. Alternatively, the same result can be achieved by the lime-soda process in which Ca^{2+} ions are precipitated as $CaCO_3$.

Salty or brackish waters can be purified by a variety of processes. The one in most common use today is the age-old process of distillation. Ion exchange columns containing two resins, one which exchanges H^+ ions for cations, the other OH^- ions for anions, can also be used. Still another approach which shows considerable promise is reverse osmosis, in which sufficient pressure is applied to the salt solution to cause water to move out of the solution through a membrane to a reservoir of pure water.

Objectives

Most of the material in this chapter lends itself to discussion and informed speculation rather than formal problem-solving. You should be able to explain, in terms that a nonscientist can understand:

1. the effect of the unusual properties of water upon the nature of our environment;

2. the peculiar molecular structures of ice and liquid water, relating these structures to physical properties;

3. the various kinds of particle interactions that arise when an electrolyte is added to water;

4. how the B.O.D. of water is measured and what it signifies;

5. what is meant by thermal pollution and what its effects are;

6. the effects of detergents and pesticides on the environment;

7. the sources and effects of pollution by heavy metals such as mercury;

8. the purpose of each of the processes used to purify municipal water supplies;

9. the advantages and disadvantages of the various processes under study for desalination.

You should also be able to calculate:

10. the mode of ionization of an electrolyte from data on colligative properties (Problem 11.9);

11. the B.O.D. of a water supply from oxygen analysis data;

12. the amounts of $Ca(OH)_2$ and Na_2CO_3 that should be added to soften water containing known concentrations of Ca^{2+} and HCO_3^- (Problem 11.14).

SELF-TEST

True or False

1. The difference in strength between the hydrogen bonds in H_2O and those in HF explains why the solid phases of these compounds have very different structures. ( )

2. The maximum density of water at $4°C$ can be explained by assuming that the molecular structure of ice does not completely disappear when it melts. ( )

3. A 0.1 m solution of KCl should freeze at about $-0.19°C$. ( )

4. It is believed that the $(C_2H_5)_4N^+$ ion promotes the formation of "flickering clusters" of water molecules. Consequently, one would expect the maximum density of solutions containing this ion to be above $4°C$. ( )

5. A high B.O.D. suggests that a water supply is contaminated with organic wastes. ( )

6. Branched-chain detergents are more readily broken down in nature than those with straight chains. ( )

7. In the lime-soda process of water softening, Ca^{2+} ions in the ( ). hard water are replaced eventually by Na^+ ions.

8. A zeolite column used to soften water could be regenerated ( ) by flushing with a concentrated solution of $CaCl_2$.

9. Salt water can be converted to pure water by the process of ( ) osmosis.

Multiple Choice

10. The orientation of covalent and hydrogen bonds about an ( ) oxygen atom in ice is best described as
    a) bent   b) planar   c) tetrahedral   d) hexagonal

11. The property of water that is most critical to the regulation ( ) of body temperature is
    a) high specific heat   b) high boiling point
    c) high heat of vaporization   d) high surface tension

12. At low concentrations, i (Equation 11.2) for $Fe(NO_3)_3$ ( ) approaches
    a) 1   b) 2   c) 3   d) 4

13. For a 0.1 m solution of KNO_3, i would probably be ( )
    a) about 1   b) slightly less than 2
    c) slightly greater than 2   d) about 5

14. An anion which promotes the growth of algae in water is ( )
    a) Cl^- b) SO_4^{2-} c) PO_4^{3-} d) CN^-

15. The principal advantage of an insecticide like Sevin over DDT ( ) is that
    a) it is less expensive
    b) it is less toxic to humans
    c) it remains effective longer
    d) it breaks down more quickly

16. Mercury-containing organic compounds resemble DDT in ( ) that they
    a) have similar toxicities
    b) usually arise from the same source
    c) accumulate in the food chain
    d) have similar molecular structures

17. Which one of the following processes does not remove   (   )
suspended matter from water?
    a) sedimentation   b) coagulation
    c) filtration   d) chlorination

18. One way to remove the unpleasant taste of chlorinated water   (   )
is to
    a) filter it   b) add NH_3 c) add NaOH   d) soften it

19. A certain water supply contains Ca^{2+} and HCO_3^- in a 1:2   (   )
mole ratio. To soften this water by the lime-soda process, one should
add
    a) only lime b) only soda c) both lime and soda d) CO_2

20. In reverse osmosis, the relationship between the applied   (   )
pressure, p, and the osmotic pressure, π, would most likely be
    a) $p < \pi$ b) $p = \pi$ c) $p > \pi$ d) $p \gg \pi$

SELF-TEST ANSWERS

  1. **F**   (Tetrahedral coordination is key factor.)
  2. **T**
  3. **F**   (Closer to $-0.37°C$.)
  4. **T**   (Clusters break down more slowly as T rises.)
  5. **T**
  6. **T**
  7. **T**
  8. **F**   (Use NaCl.)
  9. **F**   (Reverse osmosis.)
10. c
11. c
12. d
13. b
14. c
15. d
16. c
17. d
18. b
19. a
20. c   (If $p \gg \pi$, membrane might break.)

SELECTED READINGS

Arrhenius, S., Über die Dissociation der in Wasser gelösten Stoffe, *Zeitschrift für physikalische Chemie,* Vol. I, pp. 631-648 (1887).
  Exercise your scientific German by reading Arrhenius' theory of electrolytic solutions. Begins with a brief summary of earlier work done by Clausius and van't Hoff.
Cleaning Our Environment: The Chemical Basis for Action, Washington, D. C., American Chemical Society, 1969.
  A survey of problems and programs and some of the chemistry involved in pollution. A useful handbook to start with.
Derjaguin, B. V., Superdense Water, *Scientific American* (November 1970), pp. 52-64.
  A discussion of the controversial "anomalous" water of Problem 11.1.
Frank, H. S., The Structure of Ordinary Water, *Science* (August 14, 1970), pp. 635-641.
  A generally readable elaboration of the "flickering clusters" model discussed in this chapter.
Goldwater, L. J., Mercury in the Environment, *Scientific American* (May 1971), pp. 15-21.
  A calm, balanced analysis of the general topic — mercury in the environment: sources and effects.
Montague, K. and P., *Mercury,* San Francisco, Sierra Club, 1971.
  A rather biased discussion written in a popular, journalistic style. Worth contrasting with the above article.
Runnels, L. K., Ice, *Scientific American* (December 1966), pp. 118-126.
  A surprising number of great scientists have worked on the problem of the structure of ice and its properties (as well as those of liquid water). This is an excellent article on this work: thorough and reasonably up-to-date.
Stoker, H. S. and S. L. Seager, *Environmental Chemistry: Air and Water Pollution,* Glenview, Ill., Scott, Foresman, 1972.
  Lots more facts and figures here than almost anywhere else. A lot of chemistry applied in a straightforward manner. No obvious bias.
Turk, A. and others, *Ecology, Pollution, Environment,* Philadelphia, W. B. Saunders, 1972.
  A very readable introduction to ecological problems of pollution. There isn't very much chemistry applied here; the problems are often not as simple to analyze as the authors suggest.

12 SPONTANEITY OF REACTION; △G AND △S

QUESTIONS TO GUIDE YOUR STUDY

1. What do you mean by reaction *spontaneity?* (What is commonly meant by the word spontaneous?) What special sense is implied here?

2. Can you decide whether or not a particular reaction will occur without even carrying out the reaction? (Recall being able to calculate heat transfer for such a case.)

3. What physical meaning do you associate with ΔG and ΔS? What kinds of measurements can you make to determine their values?

4. How are ΔG and ΔS (and ΔH) related to the masses of substances taking part in a reaction?

5. How are these quantities related to each other (and to ΔH)? What can be said about the interrelation (interconversion?) of various forms of energy? How is each of these related to the properties of atoms and molecules, like bond energy and molecular geometry?

6. How do ΔG and ΔS (and ΔH) depend on reaction conditions such as temperature and pressure? Can you predict the sign or magnitude of the effect of a change in conditions for ΔG; ΔS; and ΔH?

7. Can you predict the sign or relative size of ΔS for a given reaction? Likewise for ΔG?

8. What happens at the molecular level when the entropy of a system increases; when free energy decreases? (What is the sign convention here? Recall that a positive ΔH means that heat is transferred *to* the system.)

9. To what kinds of systems or reaction conditions are the principles of spontaneity discussed in this chapter applicable; not applicable?

10. Are you now able to decide the conditions under which any given reaction may occur? (Of what significance or usefulness is all this to the nonchemist or to society?)

11.

12.

YOU WILL NEED TO KNOW

Concepts

1. How to interpret the sign of ΔH; and, in general, how to write and interpret "thermochemical equations" — Chapter 4

Math

1. How to calculate ΔH for any reaction — Chapter 4

CHAPTER SUMMARY—OBJECTIVES

If you were to ask several nonscientists to define a "spontaneous" process, you might get a variety of answers. One response might be, "a process taking place by itself without anyone having to work to bring it about." This statement resembles the definition of spontaneity presented in this chapter, where we say that (at constant temperature and pressure) *a spontaneous process is one which is capable of producing work*. Notice that it is the inherent capacity to do work that characterizes a spontaneous reaction. We say that the combustion of methane at 25°C and 1 atm is spontaneous because this reaction, if carried out in an appropriate machine, will produce energy in the form of work. The fact that under normal conditions the energy released when methane burns is dissipated as heat does not alter our conclusion.

The capacity of a reaction to produce work can be related to a fundamental property of substances known as free energy. The difference in free energy, ΔG, between products and reactants is a direct measure of the maximum amount of work that can be obtained from a reaction. Spontaneous reactions are ones for which the products have a lower free energy than the reactants (ΔG < 0). If the free energy of the products is greater than that of the reactants (ΔG > 0), work must be done to make the reaction go and we say that it is nonspontaneous. If, perchance, the free energies of reactants and products are equal ($\Delta G = 0$), the reaction system is balanced on a knife-edge; a tiny "push" will cause it to go in one direction or the other.

We can think of the free energy change as being made up of two components; according to the Gibbs-Helmholtz equation:

$$\Delta G = \Delta H - T\Delta S$$

The enthalpy change, ΔH, which we became acquainted with in Chapter 4, represents the amount of heat absorbed or evolved when a reaction is carried out at constant pressure. If the bonds in the product molecules are stronger than those in the reactants, ΔH will be negative and the reaction will be exothermic. The Gibbs-Helmholtz equation tells us that a negative value of ΔH will tend to make ΔG negative as well. We conclude that, other things being equal, exothermic reactions will tend to be spontaneous.

The other quantity appearing in this equation, ΔS, represents the difference in entropy between products and reactants. Entropy is a measure of randomness. The entropy of a solution exceeds that of the pure components; gases have greater entropies than liquids or solids. We note from the Gibbs-Helmholtz equation that a positive value of ΔS will tend to make ΔG negative and hence contribute to spontaneity. This analysis confirms what experience tells us; order degenerates into chaos without any help from anyone.

The balance between these two factors depends upon the magnitude of the absolute temperature, T. At low temperatures, $T\Delta S$ will be small and the sign of ΔG will be that of ΔH. As temperature rises, the entropy factor plays a more important role and eventually dominates. If, as most often happens, ΔH and ΔS have the same sign, the direction in which a reaction proceeds spontaneously will reverse at some temperature. A simple example is the vaporization of water at 1 atm pressure (ΔH and ΔS both positive). Below $100°C$, ΔH is greater than $T\Delta S$, ΔG is positive, and vaporization does not occur. At $100°C$, $\Delta H = T\Delta S$, $\Delta G = 0$, and the system is at equilibrium. Above $100°C$, ΔS predominates, $\Delta G < 0$, and water at atmospheric pressure boils spontaneously.

Of the three quantities in the Gibbs-Helmholtz equation: 1) ΔH is essentially independent of both temperature and pressure; 2) ΔS can be taken to be temperature independent, at least above room temperature, but is strongly dependent upon pressure for many reactions; and 3) ΔG is ordinarily dependent upon both temperature and pressure. Throughout this chapter, we have confined our discussion to reactions taking place at 1 atm and hence have dealt almost exclusively with $\Delta G^{1\,atm}$. This quantity is often calculated from free energy of formation of compounds.

$$\Delta G^{1\,atm} = \Sigma \Delta G_f^{1\,atm} \text{ products} - \Sigma \Delta G_f^{1\,atm} \text{ reactants}$$

Since the free energies of formation given in Table 12.2 are at $25°C$ ($298°K$), calculation of free energy changes based directly upon this table give $\Delta G^{1\,atm}$ at $298°K$. (For thermochemical data, also see the Appendix.)

Knowing $\Delta G^{1\ atm}$ at 298°K and ΔH, we can use the Gibbs-Helmholtz equation to obtain $\Delta G^{1\ atm}$ at any other temperature. To do this, we apply the equation at 298°K to obtain $\Delta S^{1\ atm}$

$$\Delta S^{1\ atm} = \frac{\Delta H - \Delta G^{1\,atm}\ (at\ 298°K)}{298}$$

Then, having both $\Delta S^{1\ atm}$ and ΔH, we simply substitute the appropriate value of T and obtain $\Delta G^{1\ atm}$ at that temperature. To find the temperature at which the reaction is at equilibrium at one atm pressure, we need only set $\Delta G^{1\ atm} = 0$ and solve the equation

$$T = \Delta H / \Delta S^{1\ atm}$$

Objectives

You will be able to work any of the problems in this chapter if you have a clear understanding of what is meant by spontaneity, free energy, enthalpy, and entropy, and if you know how to use the Gibbs-Helmholtz equation in conjunction with a table of free energies of formation. In particular, you should be able to

1. express in your own words what is meant by reaction spontaneity and how it is related to the maximum work that can be obtained from a reaction and the free energy change, ΔG;

2. calculate $\Delta G^{1\ atm}$ at 298°K for various reactions from Table 12.2;

3. describe, both qualitatively and quantitatively, the contributions of ΔH and ΔS to reaction spontaneity;

4. predict the sign of ΔS for a variety of physical and chemical changes, particularly those involving gases;

5. use the Gibbs-Helmholtz equation in conjunction with Tables 4.1 and 12.2 to calculate, for a variety of reactions: $\Delta S^{1\ atm}$, $\Delta G^{1\ atm}$ at any T, and the temperature at which $\Delta G^{1\ atm} = 0$.

SELF-TEST

True or False

1. At 25°C and 1 atmosphere pressure, the reaction between N_2 ( ) and O_2 is spontaneous.

2. For the decomposition of water to the elements at 25°C and   ( ) 1 atm, $\Delta G = +56.7$ kcal. This means that at least 56.7 kcal of work has to be supplied to make this reaction go.

3. If ΔG for a reaction is positive, it is impossible to carry out ( )
the reaction unless either the temperature or the pressure is changed.

4. "Work", as the term is used in this chapter, includes radiant ( )
energy.

5. Exothermic reactions always become spontaneous as we ( )
approach absolute zero.

6. Free energies of formation of compounds are positive ( )
numbers in most cases.

7. The free energies of formation of Cu_2O and CuO at 25°C and ( )
1 atm are −35 and −30 kcal/mole respectively. This means that
Cu_2O, exposed to oxygen at room temperature, will convert
spontaneously to CuO.

8. For the reaction: PCl_5 (g) $\rightarrow$ PCl_3 (g) + Cl_2 (g), ΔS is positive. ( )

9. Automobiles increase in entropy with age. ( )

10. If ΔH and ΔS for a reaction are both negative, we expect ΔG ( )
to be negative at all temperatures.

11. Reactions for which ΔH and ΔS have the same sign will tend ( )
to reverse at high temperatures.

12. For the process CO_2 (s) $\rightarrow$ CO_2 (g), we expect both ΔH and ΔS ( )
to be positive.

13. For the process referred to in (12), ΔS should decrease with ( )
increasing pressure.

Multiple Choice

14. For the reaction at 25°C and 1 atm: ( )
$$CH_4 \text{ (g)} + 2O_2 \text{ (g)} \rightarrow CO_2 \text{ (g)} + 2 H_2 O(l),$$
which one of the following statements is *not* true?
  a) the reaction is exothermic
  b) the reaction is spontaneous
  c) $\Delta G < 0$
  d) work has to be done to make the reaction go

15. Which one of the following quantities is independent of ( )
pressure?
  a) ΔG b) ΔH c) ΔS d) W_m

16. The free energy of formation of AgCl is −26.2 kcal/mole. ΔG ( )
for the reaction: 2 AgCl(s) $\rightarrow$ 2 Ag(s) + Cl_2 (g) is:
  a) −52.4 kcal   b) −26.2 kcal   c) +26.2 kcal   d) +52.4 kcal

17. Which one of the following statements best describes the ( )
relationship between ΔG and temperature?
 a) ΔG is independent of T
 b) ΔG varies with T
 c) ΔG is a linear function of T
 d) ΔG usually decreases with T

18. Which of the following would you expect to have the lowest ( )
entropy per mole?
 a) $Li(s)$ b) $Li(g)$ c) $LiCl(s)$ d) $Cl_2(s)$

19. The heat of fusion of benzene is 2.55 kcal/mole. Its melting ( )
point is 5°C. ΔS for the melting of benzene, in cal/mole°K, is about
 a) 0.5   b) 2.6   c) 9   d) 13

20. For which of the following would you expect ΔS to be ( )
nearest zero?
 a) $C(s) + O_2(g) \rightarrow CO_2(g)$
 b) $2 SO_2(g) + O_2(g) \rightarrow 2 SO_3(g)$
 c) $CaSO_4(s) + 2 H_2O(l) \rightarrow CaSO_4 \cdot 2 H_2O(s)$
 d) $CO(g) + \frac{1}{2} O_2(g) \rightarrow CO_2(g)$

21. For a certain reaction, ΔH is 20 kcal and ΔG at 25°C is 14 ( )
kcal. ΔS in cal/mole°K is approximately:
 a) +6000   b) −6000   c) +20   d) −20

22. Vaporization is an example of a process for which: ( )
 a) ΔH, ΔS and ΔG are positive at all temperatures
 b) ΔH and ΔS are positive
 c) ΔG is negative at low T, positive at high T
 d) ΔH is strongly pressure dependent

23. For a certain reaction, ΔH = +2.5 kcal, $\Delta S^{1 \text{ atm}}$ = 10 cal/ ( )
mole°K. This reaction will be at equilibrium at 1 atm at about
 a) 0.25°C   b) 25°C   c) −23°C   d) cannot tell

24. The free energy of formation of CO is −32.8 kcal/mole at ( )
25°C; its heat of formation is −26.4 kcal/mole. As the temperature is
increased, ΔG_f will
 a) remain unchanged
 b) go through zero
 c) become less negative
 d) become more negative

25. When aniline dissolves in hexane, ΔS is positive. Why then is ( )
aniline only slightly soluble in hexane at room temperature? Choose
the best thermodynamic explanation.
   a) an equilibrium is reached, at which point, $\Delta G = 0$
   b) ΔH is positive
   c) the intermolecular forces in the two liquids are quite
      different
   d) the formation of a solution is always a spontaneous
      process

SELF-TEST ANSWERS

1. F (This is implied by the existence of the two gases in the atmos-
       phere, and proved by the fact that ΔG_f of NO is positive.)
2. T
3. F (See question 2; nonspontaneous does not mean impossible)
4. T (Radiant energy brings about the nonspontaneous process of
       photosynthesis.)
5. T
6. F
7. T (Write a balanced equation and calculate ΔG.)
8. T
9. T (At least mine does.)
10. F
11. T
12. T
13. T (Entropy of gas will decrease, solid will remain about the same.)
14. d
15. b
16. d
17. c (More descriptive than b.)
18. a
19. c
20. a
21. c
22. b
23. c
24. d (ΔS is positive.)
25. b

SELECTED READINGS

Campbell, J. A., *Why Do Chemical Reactions Occur?*, Englewood Cliffs, N. J., Prentice-Hall, 1965.

Rather light, enjoyable reading — combines the topics of this and several other chapters (including bonding and reaction rates and mechanisms) but should be understandable.

MacWood, G. E. and F. H. Verhoek, How Can You Tell Whether a Reaction Will Occur?, *J. Chem. Ed.* (July 1961), pp. 334-337.

A fairly concise statement of thermodynamic feasibility. (Note the use of the symbol F for Gibbs free energy.)

Miller, G. T. Jr., *Energetics, Kinetics, and Life: An Ecological Approach*, Belmont, Cal., Wadsworth, 1971.

A popular approach to an intuitive feel for reaction dynamics and its implications. Read it for its "thermodynamic ethic" and for its humor and human concern.

Porter, G., The Laws of Disorder, *Chemistry* (May 1968), pp. 23-25.

The first of a series of articles on the energetics and dynamics of reactions. Very readable.

Sanderson, R. T., Principles of Chemical Reaction, *J. Chem. Ed.* (January 1964), pp. 13-22.

Numerous principles are set forth (like rules) to guide the reader in predicting reaction spontaneity. Many illustrative examples are given.

Also see the readings listed in Chapter 4, particularly those of Mahan, of Pimentel, and of Strong.

13 CHEMICAL EQUILIBRIUM IN GASEOUS SYSTEMS

QUESTIONS TO GUIDE YOUR STUDY

1. What does the word *equilibrium* generally mean? What, for example, is an *equilibrium vapor pressure*?

2. How would you experimentally show that a given system is in a state of equilibrium?

3. If you could watch the molecules in an equilibrium system, what would you expect to see?

4. The preceding chapter established a thermodynamic criterion for recognizing equilibrium. What is it?

5. How do you interpret equilibrium in terms of the "driving forces" behind changes in chemical systems, ΔH and ΔS?

6. How can you predict the conditions under which equilibrium may exist? For example, at what temperature and pressure are ice and water in equilibrium?

7. Can you predict the effect of changes in conditions on a system already at equilibrium? (Recall that you have been able to predict whether or not a given reaction may spontaneously occur.)

8. Is there a general approach to describing equilibrium systems? (Chapter 9 dealt with phase equilibria; chapters 16, 17, 19 and 21 deal with several types of equilibria in water solution.)

9. Can you describe a common example of a reaction that only partially converts reactants to products?

10. If many reactions do not go to completion, what does it mean about how you are to interpret chemical equations?

11. How might you experimentally show that a given reaction is reversible?

12.

13.

YOU WILL NEED TO KNOW

Concepts

1. How to interpret the signs of ΔH and ΔG — Chapters 4, 12

Math

1. How to work problems involving molar concentration units — Chapter 10
2. How to calculate ΔH and ΔG for any reaction — Chapters 4, 12
3. How to solve quadratic equations, using the quadratic formula; how to calculate square roots (e.g., by use of slide rule or logs) — Readings listed in Preface

$$\text{(For the equation:} \quad ax^2 + bx + c = 0,$$
$$x = \frac{-b \pm \sqrt{b^2 - 4\,ac}}{2a}$$

Note that only one of the two possible solutions for x will make any physical sense.)
4. How to calculate logs and antilogs — Appendix 4

CHAPTER SUMMARY—OBJECTIVES

We have seen (Chapter 12) that a closed system will spontaneously undergo a change at constant temperature and pressure (i.e., a reaction will occur) if by so doing its free energy can decrease. That is, the free energy of a system moves toward a minimum value. What happens once the free energy is at this minimum value, even though, as is often the case, reactants have not yet been completely consumed? Nothing that you can see! Reaction ceases; the system, unless disturbed, is in a state of *dynamic equilibrium*. (Dynamic, because our notion that molecules are always going about their business is supported by experiment.) This state of apparent rest can be considered the result of two driving forces striking a balance: the tendency of a system to move toward a state of minimum energy; the tendency toward maximum entropy ($\Delta H - T\Delta S = 0$). Another interpretation of the equilibrium state considers that the rates of competing reactions have become equal: no net, observable change occurs as a result (Chapter 14).

The extent of a reaction for a system at equilibrium can be described by a certain ratio of concentrations. This ratio, the *equilibrium constant* K_c, is found to have a value which depends only on the temperature (much as the vapor pressure of a pure substance depends only on the temperature). Its value does not depend, for example, on the concentrations of reactants or products before equilibrium is reached, or on how the equilibrium state is approached.

K_c has the same form for all equilibrium systems. For the general reaction of gaseous or dissolved species, A, B, C, and D: $aA + bB \rightleftharpoons cC + dD$

$$K_c = \frac{[C]^c \, [D]^d}{[A]^a \, [B]^b},$$

where the brackets refer to molar concentration at equilibrium. This expression is commonly referred to as the Law of Chemical Equilibrium. Like all the laws we have encountered, it is only an approximation to real behavior. The more nearly gases behave like ideal gases, the more nearly does this ratio have a constant value at a given temperature. In discussing equilibria in water solutions, we will see that the law holds very well only for very dilute solutions.

We can decide whether a given system is already in a state of equilibrium, or, if not, how it will approach equilibrium. We need only to compare the given concentrations to the ratio K_c at this same temperature. (Comparison to K_c allows us to predict the change, if any, that is spontaneous, just as we have seen whether or not the free energy could decrease.)

Also by comparison to K_c, we can often predict how a given change in conditions will affect a system initially at equilibrium. The same prediction can be made qualitatively by use of *Le Chatelier's principle*: When an equilibrium system is subjected to a change in conditions, a reaction will occur so as to minimize the effect of the change. (The system will again move toward a state of minimum free energy.)

Nothing yet can be said about how fast equilibrium is approached: neither ΔG nor K_c tell us anything about reaction rates. We will consider this topic in Chapter 14.

Objectives

Before leaving this chapter, you should be able to:

1. calculate K_c from equilibrium concentrations of all species; from initial concentrations of all species and the equilibrium concentration of one;

2. use K_c to calculate the equilibrium concentration of one species, given those of all others; the equilibrium concentrations of all species, given their initial concentrations;

3. use K_c to determine the direction in which a system will move to reach equilibrium;

4. calculate K_c at one temperature, given its value at another;

5. interpret equilibrium in terms of the thermodynamic driving forces;

6. describe an equilibrium system from an atomic-molecular point of view.

SELF-TEST

True or False

1. A chemical equation describes the relative changes in the numbers of moles of reactants and products as equilibrium is approached. ( )

2. Equilibrium can be considered as a balance struck between two opposing tendencies: a system will tend to move toward a state of maximum energy; a system will tend to move toward a state of minimum entropy. ( )

3. The pressure exerted by a sample of gaseous benzene in equilibrium with liquid benzene will depend on the container volume. ( )

4. Higher temperatures would favor the production of more product in the following system: Benzene(l) $\rightleftharpoons$ Benzene(g). ( )

5. The expression for K_c always shows all gaseous or dissolved species, never pure solid or liquid species. ( )

6. Gaseous hydrogen and oxygen are allowed to react to form liquid water. The value of K_c for the reaction: $2 H_2 (g) + O_2 (g) \rightleftharpoons 2 H_2 O(l)$ will depend upon the initial relative amounts of H_2 and O_2. ( )

7. One mole of HI(g) and one mole of H_2 (g) are placed in an evacuated container at 100°C and allowed to come to equilibrium. As equilibrium is approached, one can be sure that the H_2 concentration will increase. (The reaction: $2 HI(g) \rightleftharpoons H_2 (g) + I_2 (g)$) ( )

8. The value of K_c is expected to increase with temperature for any reaction that has a negative value of ΔH. ( )

Multiple Choice

9. The expression for K_c for the equilibrium: $C(s) + CO_2(g) \rightleftharpoons$ ( )
2 $CO(g)$ is

(a) $\dfrac{2\,[CO]}{[C]\,[CO_2]}$ (b) $\dfrac{2\,[CO]}{[CO_2]}$

(c) $\dfrac{[CO]^2}{[CO_2]}$ (d) $\dfrac{[CO]}{[CO_2]}$

10. Approximately stoichiometric amounts of two reactants are ( )
mixed in a suitable container. Given sufficient time, the reactants
may be converted almost entirely to products if
   (a) K_c is much less than one
   (b) K_c is much larger than one
   (c) the free energy change is zero
   (d) the free energy change is a large positive number

11. At a certain temperature, $K_c = 1$ for the reaction: ( )
2 $HCl(g) \rightleftharpoons H_2(g) + Cl_2(g)$. In this system, then, one can be sure
that
   (a) $[HCl] = [H_2] = [Cl_2] = 1$
   (b) $[H_2] = [Cl_2]$
   (c) $[HCl] = 2 \times [H_2]$
   (d) $\dfrac{[H_2]\,[Cl_2]}{[HCl]^2} = 1$

12. Given the equilibrium constants for the following reactions: ( )

$$2\,Cu(s) + \tfrac{1}{2}\,O_2(g) \rightleftharpoons Cu_2O(s),\ K_1$$
$$Cu_2O(s) + \tfrac{1}{2}\,O_2(g) \rightleftharpoons 2\,CuO(s),\ K_2,$$

one can show that for:

$$2\,Cu(s) + \quad O_2(g) \rightleftharpoons 2\,CuO(s),\ K_c \text{ is equal to}$$

(a) $K_1 + K_2$ (b) $K_2 - K_1$
(c) $K_1 \times K_2$ (d) K_2/K_1

13. What would you predict to be the conditions that would ( )
favor maximum conversion of noxious nitric oxide and carbon
monoxide: $NO(g) + CO(g) \rightleftharpoons \tfrac{1}{2}\,N_2(g) + CO_2(g),\ \Delta H = -89.3\ kcal$?
   (a) low T, high p (b) high T, high p
   (c) low T, low p (d) high T, low p

14. The position of equilibrium would not be affected by ( )
changes in container volume for the system:
   (a) $H_2(g) + I_2(s) \rightleftharpoons 2\,HI(g)$
   (b) $N_2(g) + O_2(g) \rightleftharpoons 2\,NO(g)$
   (c) $N_2(g) + 3\,H_2(g) \rightleftharpoons 2\,NH_3(g)$
   (d) $H_2O_2(l) \rightleftharpoons H_2O(l) + \tfrac{1}{2}\,O_2(g)$

15. For the reaction: ( )
$$4 NH_3 (g) + 7 O_2 (g) \rightleftharpoons 2 N_2O_4 (g) + 6 H_2O(l),$$
increasing the pressure by the addition of neon gas would be expected
    (a) to increase the yield of N_2O_4 at equilibrium
    (b) to decrease the yield of N_2O_4 at equilibrium
    (c) to speed up the reaction in the forward direction
    (d) to make no change in the relative amounts of NH_3 and N_2O_4 at equilibrium

16. For the reaction in (15), a decrease in pressure brought about ( )
by an increase in container volume would be expected to
    (a) increase the yield of water at equilibrium
    (b) diminish the extent of the forward reaction
    (c) increase the value of K_c
    (d) have no effect on the equilibrium attained

17. Which of the following changes will invariably increase the ( )
yield of products at equilibrium?
    (a) an increase in temperature
    (b) an increase in pressure
    (c) addition of a catalyst
    (d) increasing reactant concentrations

18. In which of the following cases will the least time be required ( )
to arrive at equilibrium?
    (a) K_c is very small   (b) K_c is approximately one
    (c) K_c is very large   (d) cannot say

19. In order to reach equilibrium in a shorter time interval, which ( )
one of the following would be appropriate to most any chemical reaction?
    (a) decrease the concentrations of reacting substances
    (b) increase the temperature and pressure
    (c) decrease the temperature
    (d) use only stoichiometric amounts of reactants

20. For the following reaction, one would predict the form of K_c ( )
to be:

$$AgCl(s) + 2 NH_3 (aq) \rightleftharpoons Ag(NH_3)_2{}^+(aq) + Cl^-(aq)$$

(a) $\dfrac{[Ag(NH_3)^+{}_2] + [Cl^-]}{[NH_3]^2}$ (b) $[Ag^+] [Cl^-]$

(c) $\dfrac{[Ag(NH_3)_2{}^+] [Cl^-]}{[NH_3]^2}$ (d) $\dfrac{1}{[AgCl]}$

21. Into a one-liter flask at 400°C are placed one mole of N_2, ( )
three moles of H_2 and two moles of NH_3. If K_c for the following
reaction is about 0.5 at 400°C, what reaction, if any, can be
expected to occur?

$$N_2(g) + 3 H_2(g) \rightleftharpoons 2 NH_3(g)$$

(a) left to right   (b) right to left
(c) system is at equilibrium   (d) cannot say

22. For the reaction considered in (21), the value of K_c is about ( )
0.08 at 500°C. One can therefore say that
   (a) the reaction is endothermic
   (b) the reaction is exothermic
   (c) K_c is independent of temperature
   (d) K_c is directly proportional to the absolute temperature

23. If we start with 1.0 mole of N_2, 3.0 moles of H_2 and 2.0 ( )
moles of NH_3 in a one-liter container at 500°C (where K_c = 0.08 for
the reaction written as in 21), at equilibrium
   (a) the number of moles of N_2, H_2, and NH_3 will be in the
       ratio 1:3:2
   (b) the number of moles of N_2 and H_2 will be in the ratio
       1:3
   (c) the number of moles of N_2 will be 1.0
   (d) the total number of moles will be the same as at 400°C

24. How do the equilibrium constants for forward and reverse ( )
reactions compare to each other?
   (a) they are always the same
   (b) their sum must equal one
   (c) their product must equal one
   (d) they are not related

25. If a system contains SO_2, O_2, and SO_3 gases at equilibrium, ( )

$$SO_2(g) + \tfrac{1}{2} O_2(g) \rightleftharpoons SO_3(g)$$

the addition of more oxygen to the system will result in
   (a) a reaction in which some SO_2 is formed
   (b) a reaction in which part of the added O_2 is consumed
   (c) a reaction in which all of the added O_2 is consumed
   (d) no reaction

26. It is sometimes the case that K_c is very small for a reaction ( )
under almost all conditions. Yet the reaction may be used to produce
significant amounts of products. How can this be?

    (a) a catalyst is used to increase the yield
    (b) the reaction is carried out at very high temperatures
    (c) an alternate series of reactions, giving the same result, is
        used
    (d) product is removed from the system as it is formed

SELF-TEST ANSWERS

  1. **T** (An equation does not, for example, tell us how fast equilibrium
       is approached.)
  2. **F** (Minimum energy, maximum entropy both favor spontaneity.)
  3. **F** (Equilibrium vapor pressure depends only on temperature.)
  4. **T**
  5. **T**
  6. **F** (K_c for a given reaction depends only on temperature.)
  7. **T** (Hint: I_2 must form.)
  8. **F**
  9. **c**
10. **b**
11. **d** (The composition could be anything, provided this ratio equals
      K_c.)
12. **c** (Hint: Write out all three expressions for K_c.)
13. **a**
14. **b** (Numbers of moles of gas unchanged.)
15. **d** (Neon takes no part in the net reaction.)
16. **b** (Favoring the larger number of moles of gas.)
17. **d**
18. **d** (See Chapter 14.)
19. **b** (Again, Chapter 14.)
20. **c** (Dissolved species only!)
21. **a** (Concentration quotient is less than K_c.)
22. **b**
23. **b**
24. **c**
25. **b**
26. **d** (Removal of product shifts equilibrium to the right.)

SELECTED READINGS

Guggenheim, E. A., More about the Laws of Reaction Rates and of Equilibrium, *J. Chem. Ed.* (November 1956), pp. 544-545.
A brief discussion of the history of the law of equilibrium and errors commonly made about the law. Follows the discussion in the article by Mysels.

Mysels, K. J., The Laws of Reaction Rates and of Equilibrium, *J. Chem. Ed.* (April 1956), pp. 178-179.
A short discussion of errors commonly made in relating the law of equilibrium to rates of reactions. (Perhaps more appropriate reading after Chapter 14.)

Further readings, at the level of the text as well as somewhat below, can be found in the list for Chapter 12.

Further work in solving problems on chemical equilibrium can be taken from the problem manuals listed in the Preface.

14   RATES OF REACTION

QUESTIONS TO GUIDE YOUR STUDY

1. Knowing what reactions *can* occur (criteria have been established in Chapters 12 and 13), can you now say what reactions *will* occur? In particular, can you say when a given reaction will begin, how fast it will go, and when it will stop?

2. What properties of a system could you observe to see how fast a reaction is occurring? What kinds of measurements would you make? How can you measure the rates for very fast and very slow reactions?

3. How does the rate of a reaction depend on conditions such as temperature, pressure and the physical state of the reactants?

4. How does the rate of a reaction depend on concentration? How do you experimentally arrive at such a relationship (the *rate law*)? How do you account for a particular rate law in terms of atomic-molecular behavior?

5. What can you conclude about the nature of individual molecular rearrangements (the sum of these *elementary reactions* constitutes the *reaction mechanism*) from a study of reaction rates? Is there ever more than one possible mechanism?

6. How do you account for the fact that most reactions speed up as temperature is increased? Can you quantitatively relate rate and temperature? (Recall that kinetic theory relates molecular speeds and temperature.)

7. How does the rate of a reaction change with time? (Recall that for closed systems, a reaction is expected to approach equilibrium.)

8. How can you change or control the rate of a reaction? (Any common examples you can think of?) How are reaction rates controlled or altered in living organisms?

9. Can you predict or calculate the rate, or write the rate law, for a particular reaction without ever carrying out that reaction? (Recall that you have been able to calculate ΔH, ΔG, K_c for such a case.)

10. In what ways are rate laws useful? What, if anything, do they let you say about the feasibility of carrying out a particular reaction?

11.

12.

YOU WILL NEED TO KNOW

Concepts

1. How to interpret the sign of ΔH — Chapter 4
2. How to interpret the Maxwell distribution of molecular energies — Chapter 5

Math

1. How to calculate logs and antilogs — Appendix 4
2. How to work problems involving molar concentration units — Chapter 10
3. How to work with the graph of an equation for a straight line (e.g., $\log k = -B/T + A$ has the form $y = mx + b$) — See a math text

---

CHAPTER SUMMARY—OBJECTIVES

From the free energy change (Chapter 12) or the equilibrium constant (Chapter 13), we can determine whether or not a reaction will take place and, if so, the extent to which it will occur, *given sufficient time*. Thermodynamic quantities such as ΔG and K_c are derived from experimental measurements; they require no knowledge of molecular behavior. In contrast, when we deal with rate of reaction we try to establish the path or mechanism by which it occurs. This requires that we make some assumptions as to what the molecules are up to. Hopefully, these assumptions can be checked by experimental measurements which tell us how reactant concentrations change with time.

The simplest molecular model for reaction path assumes that in order for reaction to occur, two high energy molecules must collide. This implies that reaction rate should increase with concentration; the more molecules there are in a given volume, the more collisions there will be in unit time. Experimentally, we find that rate does indeed increase with concentration.

However, the relationship is more complicated than a simple collision mechanism would imply. If the path of every reaction involved nothing more than a simple, two-molecule collision, all reactions would be second order. That is, in the general rate expression

$$A + B \rightarrow \text{products; rate} = k(\text{conc } A)^m (\text{conc } B)^n$$

we would expect that: $m + n = 2$. Experimentally, we find that reactions of other orders are common. In particular, we frequently encounter first order reactions where the relationship between concentration and time is given by the equation:

$$\log_{10} \frac{X_0}{X} = \frac{kt}{2.30}$$

where X_0 and X are reactant concentrations at $t = 0$ and t respectively. Reactions of other integral orders (0, or 3) or fractional orders $(\frac{1}{2}, \frac{3}{2})$ are also known.

The basic weakness of the simple collision model is that very few reactions proceed by a single step. More frequently, the reaction mechanism involves a series of steps, each of which may be a bimolecular collision or the decomposition of an unstable, high-energy species. In most cases, one step is considerably slower than the others and hence determines both the overall rate and the observed reaction order. Experimental data on reaction rate may suggest what this step is but it can never establish an unambiguous mechanism for a particular reaction.

The collision theory explains quite satisfactorily the dependence of reaction rate upon temperature. If we assume that only collisions between high-energy molecules are effective, we can derive the Arrhenius relation between rate constant and temperature:

$$\log_{10} k = \text{constant} - \frac{E_a}{(2.30) RT}$$

The activation energy, E_a, represents the minimum energy that must be available if a collision is to be fruitful. We see from the Arrhenius equation that a reaction which requires a high activation energy will ordinarily be slow (small rate constant). Conversely, if E_a is very small, most of the molecules have sufficient energy to react when they collide and reaction should occur very rapidly.

It is possible to lower the activation energy by finding a different reaction path with a lower energy barrier. In the laboratory, we do this by finding an appropriate catalyst for the reaction. Frequently, a catalyst provides an active surface upon which reaction can occur. Remember that a catalyst changes only the activation energy and hence the rate of reaction; it does not affect the overall energy difference between products and reactants and hence cannot change the equilibrium constant.

Objectives

Before attempting to work problems dealing with reaction rate,

1. you should be able to define or explain what is meant by each of the following terms:

reaction rate	chain reaction
order of reaction	half-life
rate constant	activated complex
activation energy	mechanism of reaction
catalyst	

You should then be able to:

2. determine reaction order, given initial rate as a function of reactant concentrations;

3. use Equation 14.5 to calculate, for a first order reaction, from appropriate data
   a) the concentration of reactant remaining after time t
   b) the time required for the concentration to decrease to a given value
   c) the half life;

4. use Equations 14.8 and 14.9 to determine
   a) E_a, knowing k as a function of T
   b) k at one T, knowing E_a and the value of k at another T
   c) the temperature at which k will have a certain value;

5. use Table 14.4 to deduce whether a reaction is 0, 1st or 2nd order (Problem 14.23);

6. write a plausible mechanism for a simple reaction given the rate expression (Problem 14.13);

7. discuss the triumphs and deficiencies of the collision theory of reaction;

8. discuss some general features of enzyme-catalyzed and surface reactions.

SELF-TEST

True or False

1. The rate of reaction ordinarily decreases with time. ( )

2. If the rate of reaction doubles when the concentration is doubled, the reaction must be first order. ( )

3. For a zero order reaction the rate constant k could have the units sec^{-1}. ( )

4. In a first order reaction a plot of concentration vs time is a ( )
straight line.

5. In general, very fast reactions have small activation energies. ( )

6. The activation energy for a reaction can be obtained by ( )
taking the difference in energy between reactants and products.

7. An enzyme-catalyzed reaction may be inhibited by adding a ( )
substance with a structure very similar to that of the substrate.

8. Any reaction for which the rate determining step involves a ( )
collision between two molecules must be second order.

9. The order of reactions occurring at solid surfaces frequently ( )
changes with concentration.

10. It is possible for the activation energy, as calculated from the ( )
Arrhenius equation, to be negative.

11. The reaction: $H_2(g) + I_2(g) \rightarrow 2\ HI(g)$ is first order in both ( )
H_2 and I_2. Consequently, the mechanism must involve a simple
bimolecular collision between H_2 and I_2 molecules.

Multiple Choice

12. For the reaction: $2\ NO(g) + O_2(g) \rightarrow 2\ NO_2(g)$, the rate is ( )
expressed as $-\Delta conc\ O_2/\Delta t$. An equivalent expression would be
    a) $\Delta conc\ NO_2/\Delta t$ b) $-\Delta conc\ NO_2/\Delta t$
    c) $-2\Delta conc\ NO_2/\Delta t$ d) none of these

13. For a certain decomposition the rate is 0.30 mole/lit sec ( )
when the concentration of reactant is 0.20 M. If the reaction is
second order, the rate (mole/lit sec) when the concentration is 0.60
M will be:
    a) 0.30   b) 0.60   c) 0.90   d) 2.7

14. In a first order reaction the half life is 20 minutes. The rate ( )
constant k in min^{-1} is about
    a) 0.035   b) 0.35   c) 13.9   d) cannot tell

15. For a reaction of 3/2 order, it takes 20 minutes for the ( )
concentration to drop from 1.0M to 0.60 M. The time required for
the concentration to drop from 0.60 M to 0.20 M will be
    a) more than 20 min   b) 20 min
    c) less than 20 min   d) cannot tell

16. Two reactions, R_1 and R_2, have activation energies of 40   ( )
kcal and 20 kcal, respectively. Which one of the following statements
must be true?
    a) R_1 is faster than R_2 at any given T.
    b) R_1 is slower than R_2 at any given T.
    c) the rate of R_1 increases with T more rapidly than that of
       R_2.
    d) the rate of R_1 is doubled by increasing the temperature
       $10°C$.

17. The activation energy of a certain reaction is 15 kcal. The   ( )
activation energy for the reverse reaction is
    a) −15 kcal b) > 15 kcal c) < 15 kcal d) cannot tell

18. The effectiveness of a catalyst depends upon its ability to   ( )
    a) decrease the activation energy   b) increase K_c
    c) increase reactant concentration d) increase temperature

19. The principal reason for the increase in reaction rate with   ( )
temperature is:
    a) molecules collide more frequently at high temperatures
    b) the pressure exerted by reactant molecules increases with
       T
    c) the activation energy increases with T
    d) the fraction of high energy molecules increases with T

20. For the chain reaction between H_2 and F_2, the step: ( )
$H + F \rightarrow HF$ represents
    a) chain initiation
    b) chain propagation
    c) chain termination
    d) the overall mechanism of the reaction

21. The decomposition of ozone is believed to occur by the   ( )
mechanism

$$O_3 \rightleftharpoons O_2 + O$$
$$O + O_3 \rightarrow 2 O_2 \quad \text{(rate determining)}$$

When the concentration of O_2 is increased, the rate will
    a) increase b) decrease c) stay the same d) cannot say

22. Enzyme-catalyzed reactions resemble surface reactions most   ( )
closely in
    a) mechanism   b) E_a c) ΔG d) ΔH

23. In the reaction: $A \rightarrow B \rightarrow C$, the concentration of the inter- ( )
mediate B is likely to
    a) increase steadily with time
    b) decrease steadily with time
    c) be independent of time throughout the reaction
    d) remain constant through most of the reaction

24. Which of the following statements is true for all first order ( )
reactions?
    a) the activation energy is very low
    b) the concentration of reactant does not change with time
    c) the rate constant, k, is zero
    d) the rate is independent of time

25. The following mechanism is proposed for the oxidation of ( )
iodide ion

$$NO + \tfrac{1}{2} O_2 \rightarrow NO_2$$
$$NO_2 + 2\, I^- + 2\, H^+ \rightarrow NO + I_2 + H_2O$$
$$I_2 + I^- \rightarrow I_3^-$$

A catalyst in this reaction is
    a) NO   b) I^- c) O_2 d) H^+

SELF-TEST ANSWERS

    1. T
    2. T
    3. F   (Mole/lit sec.)
    4. F
    5. T
    6. F
    7. T
    8. F   (Only if the molecules that collide are the two reactants.)
    9. T
  10. T   (Some reactions slow down as T rises.)
  11. F   (Cannot deduce mechanism unambiguously from rate law.)
  12. d   ($\tfrac{1}{2} \Delta\text{conc } NO_2 / \Delta T$)
  13. d
  14. a
  15. a
  16. c
  17. d
  18. a
  19. d

20. c
21. b   (Conc O inversely related to that of O_2)
22. a
23. d   (Has to be zero at beginning and end)
24. d
25. a

SELECTED READINGS

Bunting, R. K., Periodicity in Chemical Systems, *Chemistry* (April 1972), pp. 18-20.
    The title refers to periodic recurrences during chemical reaction in certain systems of coupled reactions.
Edwards, J. O., From Stoichiometry and Rate Laws to Mechanism, *J. Chem. Ed.* (June 1968), pp. 381-385.
    How does the chemist decide on a mechanism? This article presents some of the "working rules" followed. Somewhat advanced.
King, E. L., *How Chemical Reactions Occur: An Introduction to Chemical Kinetics and Reaction Mechanisms*, New York, W. A. Benjamin, 1964.
    A good introduction to many aspects of kinetics — catalysis, very fast reactions, experimental methods — includes exercises.
Tamaru, K., New Catalysts for Old Reactions, *American Scientist* (July-August 1972), pp. 474-479.
    Partly a general review, this article presents a discussion of some new catalysts based on graphite, with unusual properties.
Wolfgang, R., Chemical Accelerators, *Scientific American* (October 1968), pp. 44-52.
    Beams of molecules passing through a vacuum find use in the new tool for studying reaction mechanisms and energies.

For further discussion of kinetics, and its connection with equilibrium, see the books by Campbell and by Miller listed in Chapter 12.

15 THE ATMOSPHERE

QUESTIONS TO GUIDE YOUR STUDY

1. What are some of the bulk properties of the atmosphere? What are its dimensions and mass? Can the ideal gas law be applied to relate some of these properties?

2. What is the overall composition of the atmosphere? How does the composition vary with weather, geographical location and altitude? What kinds of experiments give this information; what kinds of explanations do we give?

3. How may the components of the atmosphere be separated and identified?

4. What are the origins of the atmospheric components which are most abundant? (For example: What reactions are known to give rise to atmospheric oxygen; to deplete it?)

5. What reactions do the components of the atmosphere take part in, within the atmosphere itself, as well as at the interface of atmosphere and earth and between the atmosphere and living organisms?

6. What are the rates and extents of these reactions?

7. What substances can be considered as atmospheric contaminants? Where do they come from; where do they go? What effects do they have on atmospheric properties; on life and other processes?

8. How do you test for pollutants? How do you specify their concentrations? How do you control them? How do you prevent them?

9. How do you measure the extent of pollution and its change with time? (Is pollution increasing?)

10. What are some of the reactions by which pollutants can be removed? What are their limitations; their costs?

11.

12.

YOU WILL NEED TO KNOW

Concepts

1. How to interpret the sign and relative magnitudes of thermodynamic quantities (ΔH, ΔS, ΔG) and relate these to changes at the atomic-molecular level — Chapters 4, 12

Math

1. How to predict, qualitatively as well as quantitatively, the effects of changes in reaction conditions (T, P, concentration, nature of reactant, etc.) on:

   spontaneity of reaction (sign of ΔG) — Chapter 12
   equilibrium concentrations — Chapter 13
   rate of reaction — Chapter 14

2. How to work with concentration units introduced thus far — Chapter 10

3. How to define as well as work with partial pressure — Chapter 5

CHAPTER SUMMARY—OBJECTIVES

In this chapter, we have attempted to tie together the concepts of chemical kinetics, chemical equilibrium, and chemical bonding introduced in previous chapters, to discuss the chemical and physical properties of the atmosphere. Interwoven with these concepts is a considerable amount of descriptive chemistry of the major constituents of the atmosphere: nitrogen, oxygen, carbon dioxide, water vapor, and the noble gases.

Our interest in the chemistry of the very unreactive element nitrogen centers upon the process of nitrogen fixation in which N_2 molecules are converted to useful compounds. This process is carried out in nature by certain clever bacteria found in the roots of clover and other legumes. Industrially, it is accomplished by reacting nitrogen with hydrogen to form ammonia (Haber process). Ammonia, either directly or in the form of its salts (e.g., NH_4NO_3), is used as a fertilizer. It can be converted by the Ostwald process to nitric acid, which in turn is used to make fertilizers (inorganic NO_3^- salts) and explosives (organic nitro compounds).

Oxygen reacts with all but a very few elements. With most metals, the O_2 molecule is converted to the oxide ion, O^{2-}. Certain of the 1A and 2A metals normally yield peroxides (O_2^{2-}) or superoxides (O_2^-). Perhaps the most important of the reactions of oxygen with the nonmetals is that with sulfur; the sulfur dioxide produced by combustion under ordinary conditions is converted first to sulfur trioxide and then to sulfuric acid. When an element forms more than one compound with oxygen, it is ordinarily the "higher" oxide (e.g., CuO, CO_2) which is formed at low temperatures in the presence of excess oxygen. High temperatures and limited amounts of oxygen, a "reducing atmosphere," favor the lower oxide (e.g., Cu_2O, CO).

In the upper atmosphere, we find many species, such as O_3, O, O_2^+, which are unstable under ordinary laboratory conditions. These are formed by high energy radiation in the ultraviolet and far UV regions of the spectrum; very little of this radiation reaches the surface of the earth. Since the reactions that form these species are nonspontaneous from a thermodynamic point of view, we might expect them to be readily reversed. However, because the concentrations of these atoms and ions are extremely low, their rate of recombination to form stable molecules such as O_2 is quite slow.

Among the "minor" but objectionable constituents of the atmosphere are the oxides of sulfur (mostly SO_2), the oxides of nitrogen (mostly NO), carbon monoxides, hydrocarbons, and suspended particles. Even though these species are present at a level of a few parts per million or less, they pose serious problems to human health. The automobile is a major source of three of these five pollutants (NO_x, CO, hydrocarbons) and is largely responsible for smog formation, which involves reaction of NO_2 with hydrocarbons. The combustion of fuels in power plants and industry is responsible for much of the sulfur oxides and suspended solids that enter the atmosphere.

Efforts to reduce the level of air pollution have concentrated largely upon modifications of the internal combustion engine and exhaust system of the automobile. Several different approaches are under study, but none, at least at present, seem likely to reduce NO_x and hydrocarbon emissions to the level required to prevent smog formation. Perhaps the ultimate answer will be to use an energy source other than gasoline; possibilities include natural gas, electricity, and steam.

Objectives

This chapter gives you an opportunity to review concepts of reaction rate, equilibrium and bonding; many of the problems at the end of the chapter relate directly to these topics. In addition, you should be able to

1. convert concentrations in the gas phase given in mole %, volume %, weight %, partial pressure, mole/liter, ppm, or ppb to any of the other units (Table 15.2);

2. list at least five major constituents of the atmosphere and describe how they can be extracted from it;

3. discuss the chemistry of certain industrial processes, notably the production of ammonia and sulfur trioxide, from equilibrium and rate considerations;

4. write balanced equations for the reactions of N_2 and O_2 with a variety of elements;

5. describe in words how substances such as Na_2O_2, KO_2, H_2O_2, CO, CO_2, Cu_2O, CuO, and XeF_4 can be prepared;

6. define what is meant by relative humidity and make calculations such as those called for in Problem 15.12;

7. Calculate the maximum wavelength of light that will be effective in forming unstable species such as O, O_2^+, and N_2^+ from stable molecules (Problem 15.16);

8. list five major types of air pollutants, discuss their formation, their reactions in the atmosphere, and their effects on human beings and the environment;

9. discuss various approaches to "cleaning up" the internal combustion engine, pointing out advantages and drawbacks.

SELF-TEST

True or False

1. Helium is the most abundant noble gas in the atmosphere. ( )

2. Of all the gases in the atmosphere, nitrogen is the least reactive. ( )

3. The reaction of oxygen with a 1A metal may produce O^{2-}, O_2^{2-} or O_2^- ions. ( )

4. Carbon dioxide is obtained commercially by fractional distillation of air. ( )

5. When a sample of air is allowed to warm up without changing the total water content, the relative humidity usually increases. ( )

6. As one moves up in the atmosphere, the ratio (conc O) /(concO$_2$) increases. ( )

7. The temperature profile of the atmosphere given in Figure 15.5 could, at least in principle, be obtained by taking readings on a mercury-in-glass thermometer. ( )

8. The ratio of O_3 to O concentration in the upper atmosphere ( )
can be calculated knowing the concentration of O_2 and the equilibrium constant for the reaction: $O_3 \rightleftharpoons O_2 + O$.

9. In order to reduce the concentration of NO in automobile ( )
exhaust, it is desirable to increase the temperature at which the fuel
is burned.

10. The equilibrium constant for the reaction: ( )

$$CO(g) + Hem.O_2 (aq) \rightleftharpoons O_2 (g) + Hem.CO(aq)$$

is less than one.

Multiple Choice

11. In a synthetic atmosphere made up of He and O_2, the mole ( )
fraction of He is 0.80. The weight % of He is:
    a) 80   b) less than 80   c) greater than 80   d) cannot tell

12. The formula of calcium nitride is ( )
    a) CaN   b) $Ca_2 N$ c) $Ca_2 N_3$ d) $Ca_3 N_2$

13. In the compound $Fe_3 O_4$, the mole ratio of Fe^{2+} to Fe^{3+} is ( )
    a) 1:1   b) 1:2   c) 2:1   d) 3:4

14. In the Haber process for making ammonia, high pressures are ( )
used to
    a) increase the yield, leaving the rate unchanged
    b) increase the rate, leaving the yield unchanged
    c) increase the yield and the rate
    d) increase the equilibrium constant and the rate

15. Consider the reaction: $SnO(s) + \frac{1}{2} O_2(g) \rightarrow SnO_2(s)$. This ( )
reaction would be expected to
    a) become less spontaneous at high temperatures
    b) become less spontaneous at high pressures
    c) become less spontaneous at low temperatures
    d) not occur at any temperature or pressure

16. To increase the yield of SO_3 from SO_2 in the reaction with ( )
O_2, we could
    a) increase the temperature
    b) use more SO_2
    c) add a catalyst
    d) increase the pressure

17. Of the following noble gases, which reacts most readily with   (   )
fluorine?

    a) He   b) Ne   c) Kr   d) Xe

18. Dry Ice is effective in seeding clouds because   (   )

    a) CO_2 and H_2O have similar crystal structures

    b) it increases the water content of the cloud

    c) CO_2 molecules offer a nucleus for condensation

    d) upon evaportion, it lowers the temperature of the water

19. Automobile emissions are not a major source of   (   )

    a) NO_2 b) CO   c) hydrocarbons   d) SO_2

20. Which one of the following hydrocarbons would be most   (   )
likely to contribute directly to smog formation?

    a) CH_4 b) C_2H_4 c) C_3H_8 d) C_6H_6

21. Of the following fuels, which one, under normal conditions,   (   )
produces the lowest concentration of pollutants?

    a) coal   b) wood   c) natural gas   d) petroleum

22. The superoxide ion, found in KO_2, has the structure: (   )

    a) :Ö–Ö: b) :Ö=Ö: c) :Ö: d) :Ö–Ö: e) :Ö–Ö:

23. Of the species: O_2, O_2^{2-}, O_2^-, O^{2-}

    a) which are paramagnetic? _____

    b) which contain covalent bonds? _____

    c) which react with water to form H_2O_2 ? _____

    d) which react with water to form O_3 ? _____

    e) which one is most stable to thermal decomposition? _____

24. Of the species: O_2, N_2, O, N, N^+

    a) which is most abundant in the atmosphere? _____

    b) which are unstable under ordinary laboratory conditions? _____

    c) which is most difficult to form from an energy stand-
point? _____

    d) which are held together by covalent bonds? _____

25. Of the molecules: NO_2, SO_2, O_3, CO, CO_2

    a) which are major contributors to photochemical smog?_____

    b) which can be converted to strong acids?_____

    c) which one is most abundant in automobile exhaust?_____

    d) which ones are thermodynamically unstable in the presence of O_2?_____

SELF-TEST ANSWERS

1. F
2. F   (Noble gases.)
3. T
4. F
5. F
6. T
7. F   (High T at high altitudes is calculated from molecular speeds.)
8. F   (Molecules too far apart, react too slowly to establish equilibrium.)
9. F
10. F   (If it were, CO would not be poisonous.)
11. b
12. d
13. b $(1\ Fe^{2+}, 2\ Fe^{3+}\ \text{to}\ 4\ O^{2-})$
14. c
15. a
16. d   (Is b as good an answer?)
17. d
18. d
19. d
20. b
21. c
22. a
23. a) O_2, O_2^- b) O_2, O_2^{2-}, O_2^- c) O_2^{2-}, O_2^- d) none e) O^{2-}
24. a) N_2 b) O, N, N^+ c) N^+ d) O_2, N_2
25. a) NO_2, O_3, b) NO_2, SO_2 c) CO_2 d) NO_2, SO_2, O_3, CO

SELECTED READINGS

The Biosphere, San Francisco, W. H. Freeman, 1970.

The September 1970 issue of Scientific American — *with relevant materials on the substances entering and leaving the atmosphere in "cycles."*

Bowman, W. H. and R. M. Lawrence, The Cabin Atmosphere in Manned Space Vehicles, *J. Chem. Ed.* (March 1971), pp. 152-153.

Suppose you wanted to create and carry with you your very own atmosphere. This is the problem briefly discussed in this article, with not very much chemistry though.

Kellogg, W. W., and others, The Sulfur Cycle, *Science* (February 11, 1972), pp. 587-595. 587-595.

Outlines what is known, and what we need to know, about man's and nature's contribution to the sulfur compounds in the atmosphere and the oceans; with some forecasts for the future.

Kenyon, D. H. and G. Steinman, *Biochemical Predestination,* New York, McGraw-Hill, 1969.

Chapter 3 discusses evidence for the current view of the composition and origin of the primitive earth atmosphere — as part of a discussion of the chemical origins of life.

Lewis, J. S., The Atmosphere, Clouds and Surface of Venus, *American Scientist* (September-October 1971), pp. 557-566.

Couples some basic chemistry with descriptive astronomy, with an emphasis on the comparison of earth chemistry and likely (and known) chemistry of Venus.

Stoker, H. S. and S. L. Seager, *Environmental Chemistry: Air and Water Pollution,* Glenview, Ill., Scott, Foresman, 1972.

One of the best sources for quantitative material on pollution. A lot of chemistry discussed in a straightforward manner.

Chemical and Engineering News (October 2, 1967), Special Report: Chemistry and the Solid Earth.

A very broad survey of where we stand (1967) in understanding the chemistry and physics of the geosphere. (Also see the article by Bachmann referred to in Chapter 1 of the guide.)

16 PRECIPITATION REACTIONS

QUESTIONS TO GUIDE YOUR STUDY

1. What occurs during a precipitation reaction? What would you observe during such a reaction? (Can you give any common example?)

2. Why does a precipitation occur? What are the driving forces behind the reaction? What are the ions doing?

3. What substances participate in this class of reaction? Can some generalizations be made? Are there correlations with electronic structure or with the periodic table?

4. How do reaction conditions affect the spontaneity and extent of a precipitation reaction? How does precipitation depend on the concentrations of reactant species?

5. Can you predict the direction and extent for this kind of reaction? How? (How have you predicted spontaneity and extent, qualitatively and quantitatively, for other systems?)

6. Can you apply Le Chatelier's principle to determine the effect of changes in reaction conditions? Can such predictions be made quantitative?

7. How do you write and interpret chemical equations for this class of reaction?

8. What can you say about the rates of precipitation reactions? How, for example, would you expect them to depend on temperature or concentration?

9. What can you say about the macroscopic nature of the products of a precipitation (for example, the shape and size of crystals) and how it depends on reaction conditions?

10. What are some applications for this kind of reaction? Where is it normally encountered?

11.

12.

YOU WILL NEED TO KNOW

Concepts

1. The general principles of solubility, and how solubility may change with conditions (e.g., temperature) — Chapters 10, 11
2. How to write and interpret [balanced] chemical equations — Chapter 3
3. How to define and interpret an equilibrium constant — Chapter 13

Math

1. How to perform stoichiometric calculations, particularly those involving concentration units (e.g., calculating mass of reactant solute from volume and concentration of solution) — Chapters 3, 10 (and more worked examples in Chapter 16)
2. How to qualitatively and quantitatively predict the effects of changes in reactant or product concentrations on the position of equilibrium — Chapter 13
3. How to write the equilibrium constant expression for any reaction — Chapter 13

CHAPTER SUMMARY—OBJECTIVES

This is the first of six chapters devoted to a discussion of the reactions of ions and molecules in water solution. Of the various types of reactions we will consider (precipitation, acid-base, complex ion formation, oxidation-reduction), precipitation is perhaps the simplest. A precipitation reaction will occur when two electrolyte solutions are mixed if one of the possible products is insoluble. To illustrate, when solutions of $MgCl_2$ and $NaOH$ are mixed, one of the two possible products, $Mg(OH)_2$, is insoluble (what is the other possible product?), so the following reaction occurs:

$$Mg^{2+}(aq) + 2 \, OH^-(aq) \rightarrow Mg(OH)_2 \, (s)$$

Notice that the equation written for the reaction includes only those species which actually participate in it; "spectator" ions (Na^+, Cl^-) are omitted.

In order to decide when a precipitation reaction will occur, we need to know water solubilities of various electrolytes. Unfortunately, these cannot be predicted with any confidence from first principles. At the present state of electrolyte solution theory, the best we can do is to rationalize after the fact; i.e., "explain" why $Mg(OH)_2$ should be much less soluble than NaCl. We have to resort to solubility rules (based simply on observation) such as those listed in Table 16.1 to predict the spontaneity of precipitation reactions.

Table 16.1 lists those compounds which can be expected to precipitate when 0.1 M solutions of the corresponding ions are mixed. If the solutions used are more dilute, and indeed for any quantitative calculations, we must know the solubility product of the electrolyte involved. For $Mg(OH)_2$ we find experimentally that

$$K_{sp} = [Mg^{2+}] \times [OH^-]^2 = 1 \times 10^{-11}$$

This is the equilibrium constant expression for the reaction

$$Mg(OH)_2 (s) \rightleftarrows Mg^{2+}(aq) + 2\,OH^-(aq)$$

and can be manipulated in every way like the equilibrium constants of Chapter 13. In words, this equation tells us that $Mg(OH)_2$ will precipitate from any solution in which the product of the concentration of Mg^{2+} times that of OH^- squared exceeds 1×10^{-11}. Thus, $Mg(OH)_2$ is sufficiently insoluble to precipitate not only when 0.1 M solutions of Mg^{2+} and OH^- are mixed, but also with 0.01 M or even 0.001 M solutions. (Would $Mg(OH)_2$ form if the concentrations of both ions were 0.0001 M?) A knowledge of the solubility product also enables us to carry out many other practical calculations such as determining the solubility of a salt (Example 16.3) or the percentage of an ion precipitated from solution (Example 16.5).

Precipitation reactions are used for a variety of purposes in analytical chemistry, inorganic chemistry, and industry. For example, we take advantage of the water insolubility of $Mg(OH)_2$ to:

a) detect Mg^{2+} in a mixture with other cations, none of which forms an insoluble hydroxide;

b) analyze quantitatively for Mg^{2+} by weighing the $Mg(OH)_2$ produced (or the MgO formed by heating it) when an excess of OH^- is added to a solution;

c) separate Mg^{2+} from sea water (this is the ultimate source of all the magnesium metal in use today);

d) convert a soluble hydroxide (e.g., CsOH) to another soluble salt of that cation (e.g., CsCl, Cs_2SO_4) by adding the appropriate salt of Mg^{2+} (e.g., $MgCl_2$, $MgSO_4$), filtering off the precipitated $Mg(OH)_2$ and evaporating the solution remaining.

Objectives

After completing this chapter, you should be able to:

1. use a table of solubilities to decide whether a precipitation reaction will occur when 0.1 M solutions are mixed and write a net ionic equation for that reaction;

2. calculate the number of moles of solid formed, the number of moles of each ion left in solution, and the concentration of each ion, when a precipitation reaction is carried out;

3. use K_{sp} to
   a) calculate the equilibrium concentration of one ion (e.g., OH^-) given that of the other;
   b) determine whether or not a precipitate will form when two solutions of any given concentrations are mixed;
   c) calculate the solubility of the corresponding salt (you should also be able to perform the reverse calculation);
   d) determine the fraction of an ion left in solution after precipitation;

4. determine the % of an ion in a mixture from analytical data based on precipitation reactions, either gravimetric (Problem 16.13) or volumetric (Problem 16.14);

5. devise a scheme of qualitative analysis based on precipitation reactions to separate and identify a mixture of ions;

6. use precipitation reactions to prepare (at least on paper!) a variety of soluble and insoluble electrolytes (Problem 16.17).

SELF-TEST

True or False

1. $MgSO_4$ is more soluble than $BaSO_4$ because the charge (F) density of the ions is higher.

2. Using the "charge density rule," we would expect CaF_2 to be (T) less soluble than $CaCl_2$.

3. The solubility of AgCl in 0.10 M NaCl is greater than it is in (F) water.

4. The solubility of AgCl in 0.10 M $NaNO_3$ is greater than it is (F) in water.

5. Any two salts that have the same K_{sp} value will have the (T) same solubility.

6. When equal volumes of 0.1 M solutions of M^+ and X^- are mixed, a precipitate forms if K_{sp} of MX is less than 0.01. (F)

7. A precipitate will form when enough Cl^- is added to a solution 0.10 M in Pb^{2+} to make conc $Cl^- = 0.010$ M (K_{sp} $PbCl_2$ = 1.7×10^{-5}). (X)

8. A salt MNO_3 can be prepared by a precipitation reaction which involves adding $AgNO_3$ to a solution of MCl. (T)

Multiple Choice

$.02 \quad .03$

$.200 \quad .100$

$BaCl_2 + Na_2SO_4 \rightarrow BaSO_4 + 2 NaCl$

$.1 M \quad .3 M \qquad X \text{ moles}$

9. When 200 ml of 0.10 M $BaCl_2$ is added to 100 ml of 0.30 M $Na_2 SO_4$, the number of moles of $BaSO_4$ precipitated is (B)
    a) 0.010   b) 0.020   c) 0.030   d) 0.20

10. 200 ml of 0.10 M $NiCl_2$ is added to 100 ml of 0.20 M NaOH. The number of moles of Ni^{2+} left in solution after precipitation is $(X) B$
    a) 0   b) 0.010   c) 0.030   d) 0.10

11. When solutions of $Pb(NO_3)_2$ and $Na_2 SO_4$ are mixed, the precipitate that forms is (A)
    a) $PbSO_4$ b) $NaNO_3$ c) $PbSO_4$ and $NaNO_3$ d) none

12. The net ionic equation for the reaction that occurs when solutions of $AgNO_3$ and $Na_2 CO_3$ are mixed is (B)
    a) $2AgNO_3 (aq) + Na_2 CO_3 (aq) \rightarrow Ag_2 CO_3 (s) + 2NaNO_3 (aq)$
    b) $2 Ag^+(aq) + CO_3^{2-}(aq) \rightarrow Ag_2 CO_3 (s)$
    c) $Na^+(aq) + NO_3^-(aq) \rightarrow NaNO_3 (s)$
    d) no reaction

13. The solubility product of $PbCO_3$ is 1×10^{-12}. In a solution in which $[CO_3^{2-}] = 0.2$ M, the equilibrium concentration of Pb^{2+} is (b)
    a) 1×10^{-12} M   b) 5×10^{-12} M
    c) 2×10^{-11} M   d) 1×10^{-6} M

14. The solubility product expression for $As_2 S_3$ is: $K_{sp} =$ (D)
    a) $[As^{3+}] \times [S^{2-}]$ b) $[As^{3+}]^2 \times [S^{2-}]$
    c) $[As^{3+}]^3 \times [S^{2-}]^2$ d) none of these

15. In order to remove 90% of the Ag^+ from a solution originally 0.10 M in Ag^+, the $[CrO_4^{2-}]$ (K_{sp} $Ag_2 CrO_4 = 1 \times 10^{-12}$) must be $(X) D$
    a) 1.1×10^{-12} b) 1×10^{-11} c) 1×10^{-10} d) 1×10^{-8}

16. For a salt of formula MX_2, the solubility, s, will be related to K_{sp} by (D)
    a) $s = K_{sp}$ b) $s^2 = K_{sp}$ c) $2s^3 = K_{sp}$ d) $4s^3 = K_{sp}$

17. CrO_4^{2-} is used as an indicator in the titration of Cl^- with Ag^+ (D) because
    a) it is yellow, whereas Cl^- is colorless
    b) Ag_2CrO_4, unlike AgCl, is soluble in water
    c) Ag_2CrO_4 precipitates before AgCl
    d) Ag_2CrO_4 precipitates only when virtually all the Cl^- has reacted

18. To separate and identify the ions in the mixture: Pb^{2+}, Cu^{2+},   (A) Mg^{2+}, one might add the reagents H_2S, HCl and NaOH. They should be added in the order
    a) HCl, H_2S, NaOH   b) H_2S, HCl, NaOH
    c) HCl, NaOH, H_2S d) NaOH, H_2S, HCl

19. To prepare pure RbCl from Rb_2SO_4, one might use a   (A) precipitation reaction in which the other reagent is
    a) $BaCl_2$ b) $BaSO_4$ c) NaCl   d) RbCl

20. Consider the anions Cl^-, SO_4^{2-}, CO_3^{2-}.

    a) Which form insoluble salts with Pb^{2+}?_____

    b) Which form insoluble salts with Ba^{2+}?_____

    c) Which decrease the water solubility of KOH?_____

    d) Which decrease the water solubility of $BaSO_4$?_____

SELF-TEST ANSWERS

1. F
2. T
3. F
4. T   (Ionic strength effect.)
5. F   (Only if they have the same type formula, e.g., MX.)
6. F   (Dilution lowers conc to 0.05 M.)
7. F   ($1 \times 10^{-5} < 1.7 \times 10^{-5}$)
8. T
9. b
10. b
11. a
12. b
13. b
14. d
15. d
16. d

17. d
18. a (Any other order would precipitate two ions at once.)
19. a
20. a) all b) $SO_4{}^{2-}$, $CO_3{}^{2-}$ c) none d) $SO_4{}^{2-}$

SELECTED READINGS

Fullman, R. L., The Growth of Crystals, *Scientific American* (March 1955), pp. 74-80.
  Most precipitation reactions involve the formation of crystalline solids. How does a crystal grow? Read this.

For practice in working problems involving solubility equilibria, see the problem manuals listed in the Preface.

17 ACIDS AND BASES

QUESTIONS TO GUIDE YOUR STUDY

1. What properties are characteristic of an acid? a base? Can you think of some common examples of each, other than "stomach acid"?

2. How are these properties explained in terms of atomic-molecular structure?

3. What happens when a substance dissolves in water (reacts with it?) to give an acidic solution? What happens when a solution forms which acts as a base? What species are present in such solutions?

4. Can you predict what species will give basic solutions; acidic solutions?

5. What is meant by the *strength* of an acid? How are differences in acid strength explained in terms of atomic-molecular structure? (Likewise for bases?)

6. How are such comparisons made quantitative?

7. How do you quantitatively describe a water solution of an acid or a base? (Recall that you have been able to specify the relative concentrations of species in gaseous systems at equilibrium; in saturated water solutions of slightly soluble salts.)

8. How would you experimentally arrive at such a description?

9. What role does water play in the formation of acidic and basic solutions? Is water necessary for the existence of an acid or base?

10. Why study acids and bases? Where do you encounter them?

11.

12.

YOU WILL NEED TO KNOW

Concepts

1. How to write Lewis structures — Chapter 7
2. How to predict molecular geometry and polarity — Chapter 7

Math

1. How to find logs and antilogs — Appendix 4
2. How to write the equilibrium constant expression for any reaction — Chapter 13
3. How to perform stoichiometric calculations, particularly those involving concentration units (e.g., calculating mass of reactant solute from volume and concentration of solution) — Chapters 3, 10 (and more worked examples in Chapter 16)
4. How to qualitatively and quantitatively predict the effects of changes in reactant or product concentrations on the position of equilibrium — Chapter 13

CHAPTER SUMMARY—OBJECTIVES

This chapter and Chapter 18 deal with the general topic of acid-base reactions. In this chapter we consider the properties of acids and bases; in Chapter 18 the reactions between them are discussed.

An acidic water solution is one in which there is an excess of H^+ ions; i.e., the concentration of H^+ is greater than that of OH^-. Since, for any water solution at $25°C$, $[H^+] \times [OH^-] = 1.0 \times 10^{-14}$, it follows that in an acidic solution, $[H^+] > 10^{-7}$ M. By the same token, a basic water solution is one in which there are excess OH^- ions; i.e., $[OH^-] > [H^+]$ or $[OH^-] > 10^{-7}$ M. Acidity or basicity can also be expressed in terms of pH, defined by the equation

$$pH = -\log_{10} [H^+]$$

A solution of pH 7 is said to be neutral; an acidic solution has a pH less than 7 while a basic solution has a pH greater than 7.

Any species which forms H^+ ions when added to water will form an acidic solution. From this point of view, the following would be classified as acids:

1) A large number of molecules containing ionizable hydrogen atoms. These may be binary compounds (e.g., $HCl \rightarrow H^+ + Cl^-$) or oxyacids

(e.g., $HNO_3 \rightarrow H^+ + NO_3^-$), in which the ionizable hydrogen atom is covalently bonded to oxygen.

2) A few anions containing ionizable hydrogen atoms, of which the HSO_4^- ion is typical: $HSO_4^- \rightarrow H^+ + SO_4^{2-}$.

3) Many, indeed most, cations. A hydrated cation such as $Zn(H_2O)_4^{2+}$ can lose a proton to give an acidic solution $(Zn(H_2O)_4^{2+} \rightarrow Zn(H_2O)_3OH^+ + H^+)$. The only cations which show no tendency to react in this way are those of the 1A metals and the large ions of group 2A.

Species which form basic solutions when added to water include:

1) The hydroxides of the 1A and 2A metals $(NaOH(s) \rightarrow Na^+ + OH^-)$.

2) Ammonia $(NH_3 + H_2O \rightarrow NH_4^+ + OH^-)$ and its organic derivatives such as CH_3NH_2.

3) Many, indeed most, anions, of which the F^- ion is typical:

$$F^- + H_2O \rightarrow HF + OH^-$$

A relatively few *strong acids* ionize completely when added to water. These include HCl, HBr, HI, HNO_3, $HClO_4$, and H_2SO_4 (1st ionization). Species which ionize only partially, called *weak acids,* are much more common. The strength of an acid is directly related to the magnitude of its equilibrium constant for ionization, Ka:

$$HX(aq) \rightleftharpoons H^+(aq) + X^-(aq); K_a = \frac{[H^+] \cdot [X^-]}{[HX]}$$

The only strong bases are the hydroxides of the 1A and 2A metals, which are virtually completely ionized in water. Ammonia and its organic derivatives are weak bases, in the sense that the reaction

$$NH_3(aq) + H_2O \rightleftharpoons NH_4^+(aq) + OH^-(aq); K_b = \frac{[NH_4^+] \times [OH^-]}{[NH_3]}$$

is incomplete. Anions derived from weak acids (e.g., F^-, CN^-, $C_2H_3O_2^-$, CO_3^{2-}) act as weak bases in water solution. The dissociation constant, K_b, of any weak base can be calculated by applying the Multiple Equilibrium Rule to show that

$$K_b \times K_a = K_w = 1.0 \times 10^{-14}$$

where K_a is the dissociation constant of the corresponding acid, often called the conjugate weak acid.

The relative strengths of different oxyacids can be estimated from structural considerations. Among a series of acids derived from the same element (e.g., $HClO_4$, $HClO_3$, $HClO_2$, $HClO$), acid strength increases with the number of oxygen atoms attached to the central atom. Among acids of the same type formula derived from different nonmetals (e.g., H_2SO_3, H_2SeO_3, H_2TeO_3), acid strength increases with the electronegativity of the central atom.

Throughout most of this chapter, we have, at least by implication, used the Arrhenius definition of acids and bases as species which upon addition to water give H^+ or OH^- ions respectively. The Brønsted-Lowry definition is somewhat broader; here an acid is taken to be a proton donor while a base is a proton acceptor. Thus, for the reaction:

$$NH_3(aq) + H_2O \rightleftharpoons NH_4^+(aq) + OH^-(aq)$$

the species NH_3 and OH^- are acting as Brønsted-Lowry bases while H_2O and NH_4^+ are acids. A still more general definition due to G. N. Lewis takes an acid to be an electron-pair acceptor and a base an electron-pair donor. Metal cations such as Zn^{2+} act as Lewis acids when they accept pairs of electrons from H_2O or NH_3 molecules to form the complex ions $Zn(H_2O)_4^{2+}$ and $Zn(NH_3)_4^{2+}$.

Objectives

After completing this chapter you should be able to
1. define, explain, and illustrate what is meant by:

pH	conjugate base
K_w, K_a, K_b	Brønsted acid
strong acid	Brønsted base
strong base	Lewis acid
weak acid	Lewis base
weak base	Law of Multiple Equilibria
conjugate acid	"5% rule"

2. calculate $[OH^-]$, given $[H^+]$, or vice versa;
3. convert $[H^+]$ to pH or vice versa;
4. use K_a or K_b to determine $[H^+]$, $[OH^-]$, pH, and percentage dissociation of a weak acid or base;
5. calculate K_b for a weak base given K_a for the conjugate weak acid;
6. classify various salt solutions as acidic, basic, or neutral;
7. arrange a series of oxyacids in order of increasing strength;
8. identify the Brønsted acid and base in a reaction;
9. identify the Lewis acid and base in a reaction;
10. identify the common strong acids and strong bases.

SELF-TEST

True or False

1. In a basic water solution, $[H^+] > 10^{-7}$.   (   )

2. It is impossible to have a solution with a negative pH. ( )

3. The pH of a solution prepared by dissolving 1×10^{-9} moles ( )
of HCl in a liter of water is 9.

4. A solution which gives off bubbles when Na_2CO_3 is added is ( )
acidic.

5. A solution which is colorless to phenolphthalein (Table 17.3) ( )
must be acidic.

6. There are more weak acids than strong acids. ( )

7. Solutions containing the CN^- ion are expected to be basic. ( )

8. HF $(K_a = 7 \times 10^{-4})$ is a stronger acid than HNO_2 $(K_a = 4 \times$ ( )
$10^{-4})$.

9. A solution of NH_4Cl is basic. ( )

10. The conjugate anions of strong acids are strong bases. ( )

Multiple Choice

11. The concentration of H^+ in a solution is 2×10^{-4} M. The OH^- ( )
is
a) 2×10^{-4} M   b) 1×10^{-10} M
c) 2×10^{-10} M   d) 5×10^{-11} M

12. The pH of the solution in Question 11 is ( )
a) 3.0   b) 3.7   c) 4.0   d) 10.3

13. The solution referred to in Questions 11 and 12 ( )
a) is acidic   b) is basic   c) is neutral   d) cannot say

14. The pH of a solution is 5.5. The concentration of H^+ is about ( )
a) 1×10^{-6} M   b) 1×10^{-5} M
c) 3×10^{-5} M   d) 3×10^{-6} M

15. A 0.10 M solution of HCl would have a pH of ( )
a) 0   b) 1.0   c) 7.0   d) 13.0

16. The pH of a 0.10 M solution of a weak acid would be ( )
a) less than 1   b) 1   c) greater than 1   d) cannot say

17. Which one of the following is *not* a strong acid? ( )
a) HCl   b) HF   c) HNO_3 d) $HClO_4$

18. Which one of the following is a strong base? ( )
a) $Al(OH)_3$ b) NH_3 c) C_2H_5OH d) NaOH

19. Which one of the following ions upon addition to water ( )
would give a basic solution?
a) NH_4^+ b) Na^+ c) $C_2H_3O_2^-$ d) NO_3^-

20. A certain weak acid is 10% dissociated in 1.0 M solution. In ( )
0.10 M solution, the percentage of dissociation would be
a) greater than 10   b) 10   c) less than 10   d) complete

21. K_b for NH_3 is 2×10^{-5}. K_a for the NH_4^+ ion is ( )
a) 2×10^{-5} b) 5×10^{-9}
c) 5×10^{-10} d) 2×10^{-19}

22. Of the following acids, which is the strongest? ( )
a) H_2TeO_4 b) H_2SeO_3 c) H_2SeO_4 d) H_2SO_4

23. In the reversible reaction: ( )
$HCO_3^-(aq) + OH^-(aq) \rightleftharpoons CO_3^{2-}(aq) + H_2O$, the Brønsted acids are
a) HCO_3^- and CO_3^{2-} b) HCO_3^- and H_2O
c) OH^- and H_2O d) OH^- and CO_3^{2-}

24. In the reaction: $BF_3 + NH_3 \rightarrow F_3B{:}NH_3$, BF_3 accepts an ( )
electron pair and acts as
a) an Arrhenius base   b) a Brønsted acid
c) a Lewis acid   d) a Lewis base

25. For the reaction: $HPO_4^{2-}(aq) + H_2O \rightarrow H_2PO_4^-(aq) + OH^-(aq)$ ( )
a) HPO_4^{2-} is an acid and OH^- its conjugate base
b) H_2O is an acid and OH^- its conjugate base
c) HPO_4^{2-} is an acid and $H_2PO_4^-$ its conjugate base
d) H_2O is an acid and HPO_4^{2-} its conjugate base

26. Consider the species: Cu^{2+}, F^-, H_2O, NH_4^+:

a) which, on addition to water, give acidic solutions? _____

b) which, on addition to water, give basic solutions? _____

c) which can act as Brønsted acids? _____

d) which can act as Brønsted bases? _____

e) which can act as Lewis acids? _____

f) which can act as Lewis bases? _____

27. Which one of the following might you expect to be an ( )
"active ingredient" in Brand X Antacid?
a) KOH   b) $SO_2(OH)_2$ c) $NaHCO_3$ d) NH_4Cl

SELF-TEST ANSWERS

1. F
2. F   (E.g., 10 M HCl.)
3. F   (Where did the OH⁻ ions come from?)
4. T   (Either that or it is supersaturated with gas!)
5. F   (Could, for example, have a pH of 7.5.)
6. T
7. T·
8. T
9. F   (NH_4^+ is acidic.)
10. F (Cl⁻, for example, is neutral.)
11. d
12. b
13. a
14. d
15. b
16. c (Incomplete dissociation.)
17. b
18. d
19. c
20. a (% dissociation greater for more dilute solution.)
21. c
22. d
23. b
24. c
25. b
26. a) Cu^{2+}, NH_4^+ b) F⁻ c) H_2O, NH_4^+ d) F⁻, H_2O
    e) Cu^{2+} f) F⁻, H_2O
27. c (Good old bicarbonate of soda.)

SELECTED READINGS

Chilton, T. H., *Strong Water: Nitric Acid, Its Sources, Methods of Manufacture, and Uses,* Cambridge, Mass., MIT Press, 1968.
  Readable discussion of the chemistry and the economics of production . . . for a very important industrial commodity (most of which is used in the preparation of fertilizers).
Mogul, P. H. and J. S. Schmuckler, Dilute Solutions of Strong Acids: The Effect of Water on pH, *Chemistry* (October 1969), pp. 14-17.
  What would be the pH of 1 × 10⁻⁹ M HCl (Question #3, Self-Test)? Water makes a contribution which must be counted when dealing with very dilute solutions of acids and bases.

Further practice in working problems can be gotten from the problem manuals listed in the Preface.

18 ACID-BASE REACTIONS

QUESTIONS TO GUIDE YOUR STUDY

1. What reactions involve the participation of an acid or a base, or both? What, for example, constitutes a *neutralization*?

2. What effect does acid or base strength have on the nature of these reactions?

3. What energy effects are associated with acid-base reactions? Can you explain them in terms of what the molecules and ions are doing?

4. How can you predict the spontaneity of an acid-base reaction? its extent?

5. How would you experimentally follow an acid-base reaction? What would you measure to determine, for example, its extent?

6. What can be said about the rates of acid-base reactions; about reactions in which an acid or a base plays the part of a catalyst?

7. What are the properties of a buffer system? How do you account for them?

8. Where do you find buffers? (Can you name one?) What are some applications?

9. What are some of the applications of acid-base reactions?

10. How does an acid-base indicator work?

11.

12.

YOU WILL NEED TO KNOW

Concepts

1. How to predict whether a given species will act as an acid or a base; how to predict relative strengths of acids and bases; and, in general, most of the concepts introduced in the preceding chapter — see Chapter 17 and this section of the study guide, Chapter 17
2. How to write net ionic equations — Chapter 16

Math

1. How to work problems involving equilibrium constants in general — See Chapter 16 for solubility product constants; Chapter 17, for dissociation and hydrolysis constants
2. For other essential math, see this section of Chapter 17.

CHAPTER SUMMARY—OBJECTIVES

The neutralization reaction:

$$H^+(aq) + OH^-(aq) \rightarrow H_2O; K = 10^{14}$$

takes place when any strong acid reacts with any strong base in water solution. If the acid is weak, it is more appropriate to write the equation as

$$HX(aq) + OH^-(aq) \rightarrow H_2O + X^-(aq)$$

since the principal species in solution before neutralization is the HX molecule rather than the H^+ ion. Equilibrium constants for the reactions of weak acids with strong bases are smaller than that for neutralization ($K = K_a \times 10^{14}$) but are ordinarily large enough to drive the reaction virtually to completion. A similar situation exists in the reaction of a weak base with a strong acid where $K = K_b \times 10^{14}$. Here, the weak base may be a molecule or ion in water solution (e.g., NH_3, CO_3^{2-}) or an anion of a water-insoluble salt:

$$BaCO_3(s) + 2 H^+(aq) \rightarrow Ba^{2+}(aq) + CO_2(g) + H_2O$$

Reactions of this type are commonly used to bring into solution salts containing the anions of weak acids (F^-, CO_3^{2-}, S^{2-}).

The concentration of an acid or base in water solution or in a solid mixture can be determined by titration, in which we measure the volume of a reagent of known concentration required to reach the equivalence point. The indicator used is ordinarily a weak acid. Ideally, we choose an indicator whose K_a is equal to the concentration of H^+ at the equivalence point. In a strong acid-strong base titration, the choice of indicator is not critical since $[H^+]$ changes rapidly near the equivalence point. For a weak acid or weak base, the pH changes much more slowly. Indeed, if an acid or base is very weak, the change in pH is so gradual that an ordinary acid-base titration cannot be carried out.

A solution containing a mixture of a weak acid and its conjugate weak base acts as a buffer in the sense that it prevents drastic change in pH when small amounts of strong acid or strong base are added. For a sodium acetate-acetic acid buffer, the reactions are:

addition of strong acid: $H^+(aq) + C_2H_3O_2^-(aq) \rightarrow HC_2H_3O_2(aq)$
addition of strong base: $OH^-(aq) + HC_2H_3O_2(aq) \rightarrow C_2H_3O_2^-(aq) + H_2O$

Ideally, we use a buffer system when the K_a of the weak acid is equal to the concentration of H^+ that we wish to maintain. Thus, we might use an $C_2H_3O_2^- - HC_2H_3O_2$ buffer ($K_a = 1.8 \times 10^{-5}$) to maintain constant pH in the range 4.5 to 5.0.

Acid-base reactions, like precipitation reactions discussed in Chapter 16, are widely used in analytical chemistry, inorganic chemistry and industry. Examples include:

a) Acid-base titrations to determine the per cent of HCO_3^- in a mixture.
b) Separation of a mixture of $BaCO_3$ and $BaSO_4$ by adding a strong acid to bring CO_3^{2-} into solution.
c) Preparation of volatile species such as CO_2 (add H^+ to $CaCO_3$) or NH_3 (add OH^- to a salt containing the NH_4^+ cation).
d) The Solvay process, by which $NaHCO_3$ and Na_2CO_3 are produced, ultimately from $CaCO_3$, NaCl and H_2O.

Objectives

Upon completion of this chapter you should be able to
1. carry out calculations on the stoichiometry of acid-base reactions, including titration data;
2. define normality and gram equivalent weight of acids and bases and relate them to molarity and gram formula weight;
3. explain buffer action and, given the composition of a buffer, calculate its pH before and after the addition of a strong acid or base;
4. explain how an indicator works and choose suitable indicators for various kinds of acid-base reactions;
5. write balanced net ionic equations for acid-base reactions;
6. apply acid-base reactions to prepare (on paper) such compounds as CO_2, $NaHCO_3$, Na_2CO_3, H_2S, NH_3, HCl;
7. determine, using the law of multiple equilibria, whether a sulfide (e.g., CuS or CoS) will be soluble in dilute acid;

8. describe the principles behind the Solvay process and write balanced equations for each step.

SELF-TEST

True or False

1. The equilibrium constant for the reaction of a weak acid with a strong base is smaller than that for a strong acid-strong base reaction. (T)

2. When 100 ml of 0.10 M $HC_2H_3O_2$ is added to 100 ml of 0.10 M NaOH, a neutral solution is produced. (F)

3. When 100 ml of 0.10 M HCl is added to 100 ml of 0.10 M NaOH, a neutral solution is formed. (T)

4. The equilibrium constant for the reaction of $C_2H_3O_2^-$ with H^+ is equal to $1/K_a$, where K_a is the dissociation constant for $HC_2H_3O_2$. (1)

5. The gram equivalent weight of acetic acid, $HC_2H_3O_2$, in an acid-base titration, is always equal to its gram molecular weight. ( )

6. At the equivalence point of the titration of a strong acid with a weak base, the pH is greater than 7. ( )

7. A mixture of 100 ml of 1.0 M HCl with 100 ml of 2.0 M $NaC_2H_3O_2$ would act as a buffer. ( )

8. A mixture of 100 ml of 1.0 M HCl with 100 ml of 1.0 M $NaC_2H_3O_2$ would serve as a buffer. ( )

9. Ammonia can be prepared by heating a solution of ammonium chloride with a strong acid. ( )

10. Sodium hydrogen carbonate could be prepared by saturating a solution of NaOH with CO_2 and evaporating. ( )

Multiple Choice

11. The equation for the reaction of a water solution of the weak acid HF with a solution of NaOH is best written as ( )
   a) $H^+(aq) + F^-(aq) \rightarrow HF(aq)$
   b) $H^+(aq) + OH^-(aq) \rightarrow H_2O$
   c) $HF(aq) + OH^-(aq) \rightarrow H_2O + F^-(aq)$
   d) $HF(aq) + NaOH(aq) \rightarrow H_2O + NaF(aq)$

12. The equation for the reaction of a water solution of ammonia ( )
with a water solution of HCl is best written as

    a) $NH_3(aq) + H_2O \rightarrow NH_4^+(aq) + OH^-(aq)$
    b) $NH_4^+(aq) \rightarrow NH_3(aq) + H^+(aq)$
    c) $NH_3(aq) + H^+(aq) \rightarrow NH_4^+(aq)$
    d) $NH_3(aq) + HCl(aq) \rightarrow NH_4^+(aq) + Cl^-(aq)$

13. The equation for the reaction of Ag_2CO_3, a water insoluble ( )
solid, with a strong acid is best written as

    a) $Ag_2CO_3(s) \rightarrow 2\,Ag^+(aq) + CO_3^{2-}(aq)$
    b) $CO_3^{2-}(aq) + 2\,H^+ \rightarrow CO_2(g) + H_2O$
    c) $Ag_2CO_3(s) \rightarrow Ag_2O(s) + CO_2(g)$
    d) $Ag_2CO_3(s) + 2\,H^+ \rightarrow 2\,Ag^+(aq) + CO_2(g) + H_2O$

$Ka = 10^{-4}$
$\dfrac{}{Kw}\; 10^{-14}$ 10^{10}

14. A certain weak acid has a dissociation constant of 1×10^{-4}. ( )
The equilibrium constant for its reaction with a strong base is:

    a) 1×10^{-4} b) 1×10^{-10} c) 1×10^{10} d) 1×10^{14}

15. When 500 ml of 0.10 M NaOH is reacted with 500 ml of 0.20 ( )
M HCl, the final concentration of H^+ is: $excess\;.04\;H^+$

    a) 0.10 M   b) 0.20 M   c) 0.050 M   d) 10^{-7} M

16. Which one of the following salts will not be soluble in strong ( )
acid?

    a) $PbCO_3$ b) CaF_2 c) $PbSO_4$ neutral   d) KCl

17. For the reaction: $H_2PO_4^-(aq) + 2\,OH^-(aq) \rightarrow PO_4^{3-}(aq)$ ( )
$+ 2\,H_2O$, the gram equivalent weight of $H_2PO_4^-$ is equal to

    a) $2 \times GFW$ b) GFW
    c) 1/2 GFW   d) that of PO_4^{3-}

18. When Na_2CO_3 takes part in an acid-base reaction, the ratio ( )
of its normality to its molarity is

    a) 1/2   b) 1   c) either 1/2 or 1   d) either 1 or 2

19. A certain buffer contains equal concentrations of X^- and HX. ( )
The K_b of X^- is 10^{-10}. The pH of the buffer is

    a) 4   b) 7   c) 10   d) 14

20. A buffer is formed by adding 500 ml of 0.20 M $HC_2H_3O_2$ to ( )
500 ml of 0.10 M $NaC_2H_3O_2$. What is the maximum amount of HCl
that can be added to this solution without exceeding the capacity of
the buffer?

    a) 0.01 mole   b) 0.05 mole   c) 0.10 mole   d) 0.20 mole

21. If one were to prepare a buffer by using HSO_3^- and SO_3^{2-}, ( )
its pH would probably be in the range of _____. (Ka HSO_3^- = 3×10^{-8})

    a) 6.5-8.5   b) 10-12   c) 2.0-4.0   d) 3 ± 1

22. $CuCl_2$ is prepared from $Cu(OH)_2$ by adding 0.10 M HCl to ( )
0.20 mole of solid $Cu(OH)_2$. How much HCl should be added?
a) 500 ml   b) 1000 ml
c) 2000 ml   d) some other volume

23. The most appropriate equation for the reaction referred to in ( )
Question 22 is:
a) $Cu(OH)_2(s) + 2 H^+(aq) \rightarrow Cu^{2+}(aq) + 2 H_2O$
b) $Cu^{2+}(aq) + 2 Cl^-(aq) \rightarrow CuCl_2(s)$
c) $H^+(aq) + OH^-(aq) \rightarrow H_2O$
d) $Cu(H_2O)_4^{2+}(aq) \rightarrow Cu(H_2O)_3(OH)^+(aq) + H^+(aq)$

24. The metal sulfides CoS, CuS and FeS have solubility products ( )
of 10^{-21}, 10^{-25} and 10^{-17}, respectively. The most soluble in acid will
be
a) CoS   b) CuS   c) FeS   d) cannot say

25. Consider the three indicators methyl red ($K_a = 10^{-5}$), litmus
($K_a = 10^{-7}$) and phenolphthalein ($K_a = 10^{-9}$). Which indicator would
you use to titrate
a) NH_3 with HCl?_____
b) NaOH with HCl?_____
c) $C_2H_3O_2^-$ with HNO_3?_____
d) $HC_2H_2O_2$ with NaOH?_____
e) HCN with NH_3?_____

SELF-TEST ANSWERS

1. T
2. F   (Solution contains $C_2H_3O_2^-$, hence is basic.)
3. T
4. T   (Reaction is reverse of dissociation of $HC_2H_3O_2$.)
5. T   (Monobasic acid.)
6. F   (Less than 7.)
7. T   (Equivalent amounts of $HC_2H_3O_2$, $C_2H_3O_2^-$.)
8. F   (No capacity to absorb H^+.)
9. F
10. T
11. c
12. c
13. d
14. c

15. c
16. c
17. c
18. d   (Depending on whether reacted to HCO_3^- or CO_2.)
19. a
20. b
21. a
22. d   (4000 ml)
23. a
24. c
25. a) MR   b) any one   c) MR   d) P   e) none would work

SELECTED READINGS

See the readings list for Chapter 17.

For practice in acid-base stoichiometry, see the problems manuals listed in the Preface.

19   COMPLEX IONS; COORDINATION COMPOUNDS

QUESTIONS TO GUIDE YOUR STUDY

1. What is a *complex ion;* a *coordination compound?* What special properties do they possess?

2. Where are you likely to encounter complex ions? What elements are most likely to form complexes? (Why don't all elements participate in the formation of complexes?)

3. What kind of experimental support do we have for the existence of complexes? Could we show that they exist in the solid state; in solution?

4. What is the nature of the bonding in these species? What geometries do you associate with various complexes?

5. How does the bonding and geometry of complexes account for their properties?

6. How do you account for the formation of a complex ion in terms of ΔG, or in terms of ΔH and ΔS?

7. How can you decide on the relative stabilities of complexes? (How can you predict that complex "A" will form and not complex "B"? How can you measure the extent to which one species is formed at the expense of another?) How is this quantitatively expressed?

8. How can you change the extent of a reaction in which a complex is formed or decomposed?

9. When a complex ion takes part in a reaction, how does the rate depend on the nature of the complex?

10. What uses have been found for complexes? Can you justify their study? (What natural products contain complex ions?)

11.

12.

YOU WILL NEED TO KNOW

Concepts

1. How to write electron configurations and draw orbital diagrams for transition metal atoms and ions — Chapter 6
2. The geometries of atomic orbitals — Chapter 6; and of hybrid atomic orbitals — Chapter 7
3. How to write Lewis structures for molecules and ions — Chapter 7
4. The concept of equilibrium; Le Chatelier's principle — Chapter 13

Math

1. How to calculate K_c; how to use K_c (and K_{sp} and K_a, K_b; see Chapters 16 and 17 for definitions, at least) to calculate equilibrium concentrations; to calculate the direction in which a system will move to reach equilibrium — Chapter 13

CHAPTER SUMMARY—OBJECTIVES

This chapter offers the opportunity for you to consider the real nature of most species in water solution. In particular, transition metal ions are generally complexed by the solvent or by some other, more strongly basic ligand (basic in the Lewis sense). You have already encountered complexes: recall that the explanation for the acidity of metal ions (Chapter 17) involved aquo complexes. And you may have directly observed how their formation will permit the separation of metal ions in qualitative analysis. Were these reasons not enough justification for studying complexes, we could also point out the existence of many biologically important compounds in which central atoms are bonded to more than their expected share. The energy-storing cytochromes are protein-encapsulated complexes of iron, with the metal ion bonded to six other atoms. Many common drugs, aspirin among them, may derive their potency from their ability to serve as chelating agents.

The very existence of complexes, particulary those of the transition metals, depends on there being low-energy, closely spaced orbitals available for donor pairs of electrons to go into. The d orbitals and the s and p orbitals of adjacent principal energy levels are thus most often used. For most of the transition metals, the d orbitals are at least partially empty and therefore available for bonding.

Crystal field theory is rather successful in predicting how the relative energies of these orbitals on the central atom change with the nature of the ligands. These orbital energy changes in turn provide an explanation for the dependence of properties such as color, behavior in a magnetic field, relative stability (measured, for example by K_c) . . . on the nature of the ligands. In general, for example, the stronger Lewis bases are predicted, and observed, to form the more stable complexes. Likewise, the stronger Lewis acids, those ions of greater charge density, form the more stable complexes.

A relatively simple and generally successful picture of the bonding in complexes is given by valence bond theory. For the complexes we have considered in this chapter, the following correlations can be made: When two bonds are formed by the central atom, that atom employs two equivalent, hybrid atomic orbitals of the type sp (such hybrids are directed along a straight line away from the central atom). When four bonds form, hybrids are either sp^3 (with tetrahedral geometry) or dsp^2 (square), depending on what orbitals are available. When six bonds are formed, six equivalent hybrids are derived from two d, one s and three p orbitals (d^2sp^3 or sp^3d^2). There are other hybrids that could be described; there are other coordination numbers (numbers of atoms bonded to the central one). The four mentioned are the most frequently encountered. Note that electron pair repulsion ideas would equally as well allow you to predict most of the geometries, knowing simply the coordination number. (See Chapter 7.)

Objectives

Before leaving this chapter, you should make sure that you are able to do the following:

1. determine the charge on a complex and on the central atom, given the formula for a coordination compound;

2. determine the coordination number for the central atom in any complex, which means that you will need to —

3. recognize some of the common ligands and chelating agents and their mode of bonding (how many bonds are formed and between what atoms);

4. predict the geometry, given the formula for a complex (the distinction between square and tetrahedral geometry must finally be based on experimental evidence — magnetic properties, the existence of isomers, etc.);

5. draw structural formulas representing possible isomers for a given complex (a molecular models kit would be helpful here);

6. describe the bonding for a given species in terms of the valence bond theory, using appropriate hybrid orbitals and drawing orbital diagrams;

7. give a crystal field description, and orbital diagram, for a given octahedral complex (distinguishing, for example, between high and low spin complexes);

8. give a qualitative explanation for color and its variation among complexes;

9. distinguish between lability and stability, and compare the stabilities of complexes by using the appropriate equilibrium constants (See the Objectives for Chapter 13 for the other things you should be able to do with these equilibrium constants.);

10. illustrate, with equations, specific uses of complex formation in qualitative and quantitative analyses.

SELF-TEST

True or False

1. Of the two elements, calcium and copper, the more likely to form coordination compounds is copper. (T)

2. Of the two ions, CN^- and SO_4^{2-}, the more likely to be found in metal complexes is the sulfate ion, the weaker base. (F)

3. The sign of the entropy change for the reaction: (F)

$$Ni^{2+}(aq) + dimethylglyoxime \rightarrow chelate(s)$$

is expected to be negative.

4. One would predict that $[Pt(NH_3)_4]Cl_2$ is more soluble in water than is $[Pt(NH_3)_2Cl_2]$. ( )

5. The smaller the value of the dissociation constant for a complex, the weaker the coordinate bonds in the complex. (F)

6. The dissociation constant for a complex ordinarily increases as the strength of the ligands as bases decreases. (T)

7. Any complex ion with a coordination number of four for the central atom has a tetrahedral structure. (F)

8. Geometric (cis-, trans-) isomerism is not observed in tetrahedral complexes. (T)

9. For a complex ion, *labile* means exactly the opposite of *stable*. (F)

10. The color often associated with complexes is best explained (T)
by the approach of valence bond theory.

crystal
field or
ligand field

Multiple Choice

11. In the $NiCl_4{}^{2-}$ ion, the total number of electrons around ( )
the Ni is
    a) 28   b) 34   c) 36   d) 38

12. The formula $PdCl_2(OH)_2{}^{2-}$ is known to represent two ( )
different ions. The hybrid orbitals occupied by the bonding electrons
are
    a) sp   b) sp^3 c) dsp^2 d) d^2sp^3

13. The hybridization of gold in $Au(NH_3)_2{}^+$ is: ( )
    a) dsp^2 b) dsp   c) sp^3 d) sp

14. The maximum number of possible geometric isomers for a ( )
complex having sp^3 hybridization would be
    a) 2   b) 3   c) 4   d) none

15. Geometric isomers would be expected for: ( )
    a) $Zn(NH_3)_4{}^{2+}$ b) $Zn(H_2O)_2(OH)_2$
    c) $Co(NH_3)_3Cl_2Br$ d) $Au(NH_3)_2{}^+$

16. Complex ions are held together by ( )
    a) marital bonds   b) municipal bonds
    c) coordinate covalent bonds   d) James bonds

17. Complex ions of coordination number six have a geometric ( )
structure that is
    a) square   b) linear   c) tetrahedral   d) octahedral

18. It is known that the sulfhydryl group, −SH, forms strong ( )
coordinate bonds to certain heavy metal ions. Which of the following
do you expect to be the best chelating agent for heavy metal ions?
    a) CH_3-SH b) $H-SH$
    c) $CH_3-S-S-CH_3$ d) $HS-CH_2-CH-CH_2-OH$
                                          SH

19. Which species in the following reaction acts as a Lewis acid? ( )

$$CuSO_4(s) + 4\,NH_3(aq) \rightarrow Cu(NH_3)_4{}^{2+}(aq) + SO_4{}^{2-}(aq)$$

    a) Cu^{2+} b) NH_3 c) $SO_4{}^{2-}$ d) $Cu(NH_3)_4{}^{2+}$

20. The solubility of AgCl in water may be increased by the ( )
addition of
    a) HCl   b) $AgNO_3$ c) H_2O d) NH_3

21. Of the two complexes, $Cu(H_2O)_4{}^{2+}$ and $Cu(NH_3)_4{}^{2+}$, the ( )
second is the more stable. This means that
   a) $Cu(NH_3)_4{}^{2+}$ would be the stronger acid
   b) ethylenediamine would replace H_2O faster than it would replace NH_3
   c) $Cu(NH_3)_4{}^{2+}$ has a smaller dissociation constant
   d) $Cu(NH_3)_4{}^{2+}$ has a larger dissociation constant

22. The electronic structure for the central atom in $Co(en)_2 Cl_2{}^+$ ( )
is:

   a) ⇅ ⇅ ⇅ | ⇅ | ⇅ | ⇅ ⇅ |
   b) ⇅ ⇅ ⇅ | ⇅ ⇅ | ⇅ | ⇅ ⇅ ⇅ |
   c) ⇅ ⇅ ⇅ ⇅ ↑ | ⇅ | ⇅ ⇅ ⇅ |
   d) ⇅ ↑ ↑ ↑ ↑ ↑ | ⇅ | ⇅ ⇅ ⇅ |

23. A compound has the empirical formula $CoCl_3 \cdot 4\,NH_3$. One ( )
mole of it yields one mole of AgCl on treatment with excess $AgNO_3$.
Ammonia is not removed by treatment with concentrated sulfuric
acid. The formula of the compound is best represented by
   a) $Co(NH_3)_4 Cl_3$ b) $[Co(NH_3)_4]Cl_3$
   c) $[Co(NH_3)_3 Cl_3]NH_3$ d) $[Co(NH_3)_4 Cl_2]Cl$

24. The dissociation constant for the complex ion $Zn(NH_3)_4{}^{2+}$ is ( )
the equilibrium constant for the reaction represented by the
equation:
   a) $Zn^{2+}(aq) + 4\,NH_3(aq) \rightleftharpoons Zn(NH_3)_4{}^{2+}(aq)$
   b) $Zn(NH_3)_4{}^{2+}(aq) + H_2O \rightleftharpoons Zn(NH_3)_3 (H_2O)^{2+}(aq) + NH_3(aq)$
   c) $Zn(NH_3)_4{}^{2+} + 2\,e^- \rightleftharpoons Zn(s) + 4\,NH_3(aq)$
   d) $Zn(NH_3)_4{}^{2+}(aq) + 4\,H_2O \rightleftharpoons Zn(H_2O)_4{}^{2+}(aq) + 4\,NH_3(aq)$

25. Of the following 1.0M solutions, which has the greatest molar ( )
entropy?
   a) NaCl   b) $CuCl_2$ c) $AlCl_3$ d) $[Co(NH_3)_5 Cl]Cl_2$

26. To account for the fact that $Fe(NO_3)_3$ dissolves in water to ( )
give an acidic solution, we might write:
   a) $H_2O \rightleftharpoons H^+(aq) + OH^-(aq)$
   b) $Fe(NO_3)_3(s) \rightleftharpoons Fe^{3+}(aq) + 3\,NO_3{}^-(aq)$
   c) $Fe^{3+}(aq) + 3\,H_2O \rightleftharpoons Fe(OH)_3(s) + 3\,H^+(aq)$
   d) $Fe(H_2O)_6{}^{3+}(aq) \rightleftharpoons Fe(H_2O)_5 (OH)^{2+}(aq) + H^+(aq)$

27. How might you experimentally show that a particular ( )
complex is square?

    a) two isomers should be isolated; they would show different lability

    b) x-ray diffraction for the complex in the solid state would reveal the geometry

    c) magnetic measurements might distinguish between dsp^2 and sp^3 orbitals

    d) all of the above

28. When $[Ni(NH_3)_4]^{2+}$ is treated with concentrated HCl, two ( )
compounds having the same formula, $Ni(NH_3)_2Cl_2$, designated I and
II are formed. Compound I can be converted to compound II by
boiling in dilute HCl. A solution of I reacts with oxalic acid,
$H_2C_2O_4$, to form $Ni(NH_3)_2(C_2O_4)$. Compound II does not react
with oxalic acid. Compound II is:

    a) the same as compound I   b) the *cis* isomer

    c) the *trans* isomer   d) tetrahedral in shape

SELF-TEST ANSWERS

  1. T   (Transition metal.)

  2. F   (The base donates the bonding electrons.)

  3. F   (Consider that several moles of water are produced. The water is in the aquo complex, usually indicated merely by (aq).)

  4. T   (The first gives ions in solution; the second is a neutral species.)

  5. F

  6. T

  7. F   (May also be square.)

  8. T

  9. F   (Labile refers to rate; stable, to equilibrium position.)

10. F   (Crystal field or ligand field theory.)

11. b   (For Ni^{2+} and four pairs of electrons from the four Cl^- ions.)

12. c

13. d

14. d

15. c

16. c

17. d   (With six vertices.)

18. d   (This would permit ring formation. How would you treat heavy metal poisoning?)

19. a

20. d   (To form a complex.)

21. c
22. b   (en is a chelating agent.)
23. d
24. d   (Choice *b* would be only the first step in the overall dissociation that is usually represented by K. Note that you generally omit the water from the equation in *d*.)
25. c   (Giving 4 moles of ions, compared with 3 for the complex.)
26. d   (No precipitate forms, as *c* would indicate.)
27. d
28. c

SELECTED READINGS

Bailar, J. C. Jr., Some Coordination Compounds in Biochemistry, *American Scientist* (1971), pp. 586-592.
  An anecdotal introduction to many interesting and important complexes.
Basolo, F., and Johnson, R. C., *Coordination Chemistry,* New York, W. A. Benjamin, 1964.
  A general extension of the material in this chapter, at about the same level. Molecular oribtal and crystal field theories are emphasized.
Cotton, F. A., Ligand Field Theory, *J. Chem. Ed.* (September 1964), pp. 466-476.
  An extensive, somewhat advanced discussion.
House, J. E. Jr., Substitution Reactions in Metal Complexes, *Chemistry* (1970), pp. 11-14.
  A brief and very readable discussion of several reaction mechanisms. (You might want to first review Chapter 14.)
Pauling, L., *The Nature of the Chemical Bond,* 3rd Ed., Ithaca, N. Y., Cornell, 1960.
  Chapters 5 and 9 discuss the valence bond approach to coordination chemistry. Won't be easy reading.
Perutz, M. F., The Hemoglobin Molecule, *Scientific American* (November 1964), pp. 64-76.
  The discussion of this very complicated molecule involving an iron complex includes some background material on x-ray diffraction. A prime example of modern structural chemistry achievements.
Schubert, J., Chelation in Medicine, *Scientific American* (May 1968), pp. 40-50.
  Relating physiological properties of a drug to its chemical properties and structure is usually a very difficult and incomplete task. Here's a good beginning.

20 OXIDATION AND REDUCTION: ELECTROCHEMICAL CELLS

QUESTIONS TO GUIDE YOUR STUDY

1. What reactions have you already encountered that can be classified as oxidation or reduction? What is commonly meant by *oxidation;* by *reduction*? (What, for example, is a reducing flame?)

2. How do you recognize what is oxidized and what is reduced in the equation for a redox reaction?

3. How do you write and interpret redox equations?

4. What occurs in an electrolytic cell? in a voltaic cell? (If you could watch the individual atoms, ions and molecules, what would you expect to see?)

5. How do fuel cells differ from other voltaic cells? Why aren't they in common use?

6. What energy effects are associated with reactions in electrochemical cells?

7. What generalizations can be made about what reactions may occur in an electrochemical cell? (How do you predict what is oxidized, what is reduced?) Are there any correlations to be made with the periodic table?

8. How is electrical energy quantitatively related to the masses of reacting species in electrochemical cells?

9. What can be said about the rate of redox reactions?

10. How big, in terms of everyday experience, are the common electrical units: coulomb, ampere, volt, watt? (For example, how large a current flows through your desk lamp?)

11.

12.

YOU WILL NEED TO KNOW

Concepts

1. How to draw Lewis structures — Chapter 7
2. How to write and interpret balanced net ionic equations — Chapter 16
3. How to interpret free energy changes and the sign of ΔG — Chapter 12

Math

1. How to work problems in stoichiometry — Chapter 3

CHAPTER SUMMARY—OBJECTIVES

The last type of aqueous solution reaction that we discuss in this text is oxidation-reduction which involves a transfer of electrons from a reducing agent such as Zn, H_2, or I^- to an oxidizing agent such as Zn^{2+}, H^+, or I_2. The species which is oxidized increases in oxidation number ($Zn \rightarrow Zn^{2+}$; O.N., $0 \rightarrow +2$); the species which is reduced decreases in oxidation number ($H^+ \rightarrow H_2$; O.N., $+1 \rightarrow 0$).

Oxidation numbers are assigned using a set of arbitrary rules. Perhaps the most important of these rules tells us that the sum of the oxidation numbers of all the atoms in a species is equal to the charge of that species. This rule can be applied to find oxidation numbers of elements in unfamiliar compounds. For example, for the species HBrO and BrO_3^- we find that the oxidation numbers of Br are:

$$HBrO: +1 + O.N.\ Br + (-2) = 0; O.N.\ Br = +1$$
$$BrO_3^-: O.N.\ Br + 3\ (-2) = -1; O.N.\ Br = +5$$

It is possible to balance oxidation-reduction (redox) equations without considering oxidation numbers by using the half-equation method outlined in the text. One advantage of this method is that it suggests a way of analyzing redox reactions which we will find valuable in Chapter 21. Once an equation is balanced, it can be used to make stoichiometric calculations of the type illustrated in Example 20.2.

Many familiar redox reactions occur in the world around us. The rusting of iron, the combustion of fuels, and most of the processes involved in human metabolism are redox reactions. All of these reactions and many others that we carry out in the general chemistry laboratory (e.g., the reaction of metals with acids) involve a spontaneous electron transfer from reducing agent to oxidizing agent. Any such reaction can be adapted, at least in principle, to produce electrical energy in a voltaic cell. Most of the cells in current use (dry cell, storage battery) consume chemicals that are too expensive to make them practical, large-scale sources of electrical energy. Fuel cells, in which the chemical energy available from the combustion of fuels is converted directly to electrical energy, would seem to offer a possible solution to the energy crisis. So far at least, progress in the development of such cells has been disappointingly slow.

Nonspontaneous redox reactions can be carried out by supplying electrical energy in an electrolytic cell. Many important metals (Na, Mg, Al) and industrial chemicals (Cl_2, NaOH) are produced in this way. To calculate the "yield" of products in an electrolytic cell from the quantity of electricity supplied, we use Faraday's Laws, the application of which is illustrated in Examples 20.3 and 20.4. As we might expect, electrochemical processes seldom give a 100% yield of the desired product; side reactions divert an appreciable fraction of the electrons passing through the cell.

Objectives

After completing this chapter you should be able to:
1. define and illustrate each of the following terms:

oxidation and reduction	fuel cells
oxidizing, reducing agents	GEW of reducing or
anode, cathode	oxidizing agent
electrolytic, voltaic cells	faraday

You should also be able to:

2. determine the oxidation number of an element in a molecule or ion, following the rules listed on p. 538 of the text;

3. balance redox equations for reactions in water solution;

4. determine the amount of a substance produced or the rate at which it is produced in an electrolytic cell;

5. determine the GEW of an oxidizing or reducing agent, either from an appropriate half equation or from cell data giving the amount of substance produced by a given amount of electricity.

6. Finally, you should be familiar with some of the more common electrolytic processes (e.g., preparation of Al, NaOH, Cl_2), and with some of the voltaic cells in common use today.

SELF-TEST

True or False

1. In a redox reaction, the oxidizing agent gains electrons. ( )

2. The lowest oxidation state of a 5A element is −3. ( )

3. Metals frequently show negative oxidation numbers. ( )

4. A redox equation balanced in acidic solution can be converted to apply in basic solution by adding the proper number of H_2O molecules to both sides. ( )

5. The purpose of the iron screen in the Downs Cell is to prevent Na^+ and Cl^- ions from coming in contact with each other. ( )

6. Cryolite is added in the Hall process for making aluminum so that the electrolysis can be carried out at a lower temperature. ( )

7. In the electrolysis of a water solution of NaCl, one mole of OH^- is produced for every mole of Cl^- consumed. ( )

8. Complexing agents such as CN^- are used in many electroplating processes to increase the concentration of metal ions. ( )

9. In the electrolysis of a solution of $Ag(S_2O_3)_2{}^{3-}$, three faradays of electricity are required to form one mole of silver. ( )

10. In any cell, electrolytic or voltaic, the cathode is the negative electrode. ( )

Multiple Choice

11. The oxidation number of Mn in the $MnO_4{}^-$ ion is ( )
    a) −2   b) +6   c) +7   d) +8

12. The oxidation number of P in H_3PO_4 is ( )
    a) −3   b) +1   c) +3   d) +5

13. Using Figure 20.1, decide which one of the following is not a reasonable formula for an oxide of chromium. ( )
    a) Cr_2O b) CrO   c) Cr_2O_3 d) CrO_3

14. When the half equation: $I_2 + e^- \rightarrow I^-$ is balanced, the coefficients of I_2, e^-, and I^- are, respectively, ( )
    a) 1, 1, 1   b) 1, 1, 2   c) 1, 2, 2   d) 1, 0, 2

15. When the half equation: $HSO_3^- + H_2O \rightarrow SO_4^{2-} + H^+ + e^-$ is ( )
balanced, the coefficients, reading from left to right, are
    a) 1, 1, 1, 1, 1   b) 1, 1, 1, 2, 2
    c) 1, 2, 1, 5, 2   d) 1, 1, 1, 3, 2

16. When the half equation in Questions 14 and 15 are ( )
combined, it is necessary to multiply the reduction half equation by
____ and the oxidation half equation by ____ before adding.
    a) 1, 1   b) 1, 2   c) 2, 1   d) 1, 3

17. In the electrolysis of molten magnesium chloride, the most ( )
appropriate equation for the anode reaction would be
    a) $Mg^{2+} + 2e^- \rightarrow Mg(s)$
    b) $2 H_2O + 2e^- \rightarrow H_2(g) + 2 OH^-(aq)$
    c) $2 Cl^- \rightarrow Cl_2(g) + 2e^-$
    d) $MgCl_2(l) \rightarrow Mg(s) + Cl_2(g)$

18. In the electrolysis of Al_2O_3, the ratio of the masses of Al and ( )
O_2 produced per hour is:
    a) less than one   b) 2:3   c) 1:1   d) greater than 1

19. In purifying copper by electrolysis, which electrode should ( )
be made of pure copper?
    a) anode   b) both   c) cathode   d) neither

20. A quantity of 20,000 coulombs is equal to how many ( )
faradays?
    a) 0.021   b) 0.21   c) 1.9×10^9 d) 3×10^{-20}

21. In the formation of chromium metal from Cr^{3+}, the number ( )
of grams of Cr (A.W. = 52) produced by one faraday would be
    a) 17   b) 52   c) 104   d) 156

22. In the reductuion of MnO_4^- to Mn^{2+}, the gram equivalent ( )
weight of MnO_4^- is ____ times its gram formula weight.
    a) 5   b) 1   c) $\frac{1}{3}$ d) $\frac{1}{5}$

23. When the $Zn\text{-}Cu^{2+}$ cell is used to produce electrical energy: ( )
    a) cations move toward the Zn electrode, anions to the Cu
    b) cations move toward Cu, anions toward Zn
    c) cations and anions move toward Zn
    d) cations and anions move toward Cu

24. When a lead storage battery is charged, lead sulfate ( )
    a) is formed at the cathode
    b) is formed at the anode
    c) is formed at both electrodes
    d) is removed from both electrodes

25. Which one of the following reactions could serve as a source ( )
of energy in a fuel cell?

    a) $H_2O(l) \rightarrow H_2(g) + \frac{1}{2} O_2(g)$
    b) $Zn(s) + Cu^{2+}(aq) \rightarrow Zn^{2+}(aq) + Cu(s)$
    c) $CO_2(g) \rightarrow C(s) + O_2(g)$
    d) $C(s) + O_2(g) \rightarrow CO_2(g)$

SELF-TEST ANSWERS

1. T
2. T   (As, for example, in completing an octet.)
3. F
4. F   (OH^-)
5. F
6. T   (The melting point is lowered.)
7. T
8. F   (To reduce conc.)
9. F   (One.)
10. F
11. c
12. d
13. a   (No +1 state.)
14. c
15. d
16. a
17. c
18. d
19. c   (Cu deposited there.)
20. b
21. a
22. d
23. b
24. d
25. d

SELECTED READINGS

Anderson, R. C., Combustion and Flame, *J. Chem. Ed.* (May 1967), pp. 248-260.
    Essentially a bibliography for this important class of redox reaction. Better yet, read
    Faraday's The Chemical History of a Candle.

Brasted, R. C., Nature's Geological Paint Pots and Pigments, *J. Chem. Ed.* (May 1971), pp. 323-324.

The author writes a column on applied and "relevant" chemistry. This article looks at some redox reactions that are indeed colorful.

Lawrence, R. M. and W. H. Bowman, Electrochemical Cells for Space Power, *J. Chem. Ed.* (June 1971) pp. 359-361.

A brief survey of some of the cells used for space power.

Weissman, E. Y., Batteries: The Workhorses of Chemical Energy Conversion, *Chemistry* (November 1972), pp. 6-11.

This is primarily a description of the characteristics of various batteries — their construction, operating reactions, uses, etc.

For practice in balancing redox equations, see the problem manuals listed in the Preface. (Barrow gives programmed exercises for the use of the "oxidation number method.")

21 OXIDATION-REDUCTION REACTIONS : SPONTANEITY AND EXTENT

QUESTIONS TO GUIDE YOUR STUDY

1. How do you know what voltage to apply in carrying out an electrolysis reaction? How do you know what voltage to expect in a voltaic cell?

2. What factors determine the potential associated with any given redox reaction? (Can you relate the potential to properties of atoms such as ionization potential and electronegativity?)

3. How can you predict the spontaneity of any given redox reaction? (How have you been able to predict the spontaneity of other reactions?)

4. What factors determine the extent to which a redox reaction proceeds? How can you change the extent of reaction?

5. What quantitative relationships exist between a cell potential and reaction conditions such as temperature, concentrations and pressure?

6. How might you experimentally show that a given redox reaction is reversible? (Can you think of any common example?)

7. How do you decide what materials may be used for electrodes in voltaic cells; in electrolytic cells?

8. Why does the voltage drop during the use of a voltaic cell? Why do batteries "run down" even when not in use?

9. Are voltaic cells practical major sources of energy? (For example, for lighting and heating a house; for running a car?)

10. What can you say about the extent to which you can convert chemical energy into electrical energy and vice versa?

11.

12.

YOU WILL NEED TO KNOW

Concepts

1. How to recognize and define oxidation and reduction; how to balance redox equations; and other concepts developed in the preceding chapter
2. How to interpret the free energy change for a reaction; how to predict the effects of changes in reaction conditions on the free energy change and on the equilibrium constant — Chapters 12, 13

Math

1. How to use logs and antilogs — Appendix 4
2. How to work stoichiometric problems for redox reactions — Chapter 20
3. How to calculate the free energy change for any reaction; the equilibrium constant for the reaction; and how to quantitatively predict the effect of changes in reaction conditions on ΔG and K — Chapters 12, 13
4. How to calculate K for "multiple equilibria" — see Chapter 17

CHAPTER SUMMARY—OBJECTIVES

From measurements on voltaic cells, it is possible to obtain numbers called *standard potentials* which are a quantitative measure of the tendency of a species to be reduced or oxidized. A large positive value for the standard reduction potential implies a species is easily reduced (e.g., $Cl_2(g) + 2 e^- \rightarrow 2 Cl^-(aq)$; S.R.P. = +1.36V). Species which are very difficult to reduce have large negative standard reduction potentials ($Al^{3+}(aq) + 3 e^- \rightarrow Al(s)$; S.R.P. = −1.66V). Standard oxidation potentials, which can be obtained by changing the sign of the potential for the reverse reaction, can be interpreted similarly. Species which are difficult to oxidize, such as Cl^-, have large negative potentials (S.O.P. Cl^- = −1.36V); large positive potentials imply a species which is readily oxidized, such as aluminum (S.O.P. = +1.66V).

The standard voltage, $E°$, corresponding to a particular redox reaction can be obtained by adding the standard potentials for the two half-reactions. If the calculated value of $E°$ is positive, we conclude that the reaction is spontaneous at standard concentrations. Such reactions will take place under ordinary laboratory conditions; alternatively, they can serve as a source of electrical energy in a voltaic cell. In contrast, a reaction with a negative $E°$ value is nonspontaneous at standard concentrations; it can be carried out only by supplying electrical energy. The electrolyses of Al_2O_3 and NaCl, discussed in Chapter 20, are examples of reactions in this category.

The $E°$ value for a reaction can be directly related to the standard free energy change and hence to the equilibrium constant for the reaction. The relation is:

$$\log_{10} K = \frac{n\, E°}{0.059}$$

where n is the number of moles of electrons transferred in the reaction. From this equation, we see that a redox reaction which has a positive $E°$ value will have an equilibrium constant greater than 1. If $E°$ is negative, K will be less than 1 and the reaction will be nonspontaneous at standard concentrations.

Frequently, we need to know the voltage, E, of a cell when reactants and/or products are present at other than standard concentrations (1 atm for gases, 1 M for species in aqueous solution). For the general redox reaction:

$$aA + bB \rightarrow cC + dD$$

we write the Nernst equation:

$$E = E° - \frac{0.059}{n} \log_{10} \frac{(\text{conc C})^c\ (\text{conc D})^d}{(\text{conc A})^a\ (\text{conc B})^b}.$$

This equation tells us that the voltage drops $(E < E°)$ if the concentration of a product is increased (e.g., conc C > 1 M); we can increase the voltage of a cell by decreasing the concentration of a product (conc C < 1 M) or increasing that of a reactant (conc A > 1 M). The Nernst equation is useful for obtaining concentrations of ions in solution from voltage measurements, particularly with components too dilute to be analyzed for by ordinary chemical methods. The pH meter works on this principle; K_{sp} values (Chapter 16) and K_w, K_a and K_b for weak acids or bases (Chapter 17) can also be obtained in this way.

In the last two sections of this chapter, we apply the principles reviewed above to organize the descriptive chemistry of redox reactions in water solution. In particular, we examine some of the more important reactions of strong oxidizing agents (species with large positive reduction potentials). Most of these species fall in one of two categories: they are either highly electronegative nonmetals (F_2, Cl_2, O_2), or oxyanions in which the central atom is in its highest oxidation state (MnO_4^-, $Cr_2O_7^{2-}$, NO_3^-). Perhaps the most important redox reaction, at least from an economic standpoint, is the corrosion of iron and steel. Research in this area indicates that corrosion occurs by an electrochemical mechanism. At an anodic area on the surface of an iron object, Fe atoms are oxidized, first to Fe^{2+} and eventually to a product with the approximate composition $Fe(OH)_3$. At the cathode of the tiny voltaic cell, dissolved oxygen is reduced to H_2O molecules or OH^- ions.

Objectives

After completing this chapter, you should be able to
1. use standard potentials (Table 21.1) to
   a) calculate $E°$ values for voltaic and electrolytic cells;
   b) compare the relative strengths of different oxidizing agents; different reducing agents;
   c) determine whether redox reactions will occur spontaneously under standard concentrations;
   d) calculate $\Delta G°$ and, hence, K for a redox reaction; use K in calculations involving the position of a redox equilibrium
2. use the Nernst equation to
   a) determine the voltage of a cell or the potential of a half cell at non-standard concentrations;
   b) determine the concentration of species in solution from measured voltages; use these concentrations to obtain K values for non-redox reactions
3. discuss the chemistry of redox reactions involving Cl_2, NO_3^-, $Cr_2O_7^{2-}$ or CrO_4^{2-}. You should be familiar with the oxidation states of the three elements involved (Cl, N, Cr) and the effect of pH on equilibria involving these oxidation states.
4. discuss the electrochemical mechanism for the corrosion of iron, pointing out its practical applications.

SELF-TEST

True or False

1. In Table 21.1, the strongest oxidizing agents are located at   (   )
the lower left of the table.

2. The standard reduction potentials of F_2 and Ag^+ are +2.87V   (   )
and +0.80V respectively. We conclude that F^- is a better reducing
agent than Ag metal.

3. For a certain mixture, positive $E°$ values are calculated for ( )
two different redox reactions. We can be confident that the one with
the higher $E°$ value will occur first.

4. A species with a large positive standard oxidation potential ( )
will be a strong oxidizing agent.

5. The voltage of a cell in which the reaction: $Cu(s) + 2\ Ag^+(aq)$ ( )
$\rightarrow Cu^{2+}(aq) + 2\ Ag(s)$ occurs will be, at standard concentrations:
S.O.P. Cu + 2 × S.R.P. Ag^+

6. Reactions which are readily reversed by a small change in ( )
concentration are ones in which $E°$ is close to zero.

7. For the cell referred to in Question 5, increasing the concen- ( )
tration of Ag^+ by a factor of 10 will increase the voltage by +0.06V.

8. The oxidizing strength of oxyanions is ordinarily greatest at ( )
low pH.

9. The stable species of the element nitrogen in the +3 ( )
oxidation state in acidic solution is HNO_3.

10. The phrase "cathodic protection" refers to the common ( )
practice of enclosing fragile metal electrodes in plexiglass to prevent
them from being broken.

Multiple Choice

11. Referring to Table 21.1, if the standard reduction potential ( )
of Ni^{2+} were set at 0.00V, that of Mg^{2+} would be
    a) −2.12V   b) +2.12V   c) −2.62V   d) +2.62V

12. The standard reduction potentials of Cl_2 and Cu^{2+} are ( )
+1.36V and +0.34V, respectively. The $E°$ value for the reaction:
$Cu^{2+}(aq) + 2\ Cl^-(aq) \rightarrow Cu(s) + Cl_2(g)$ is
    a) −2.38V   b) −1.70V   c) −1.02V   d) +1.70V

13. The $E°$ values for the following reactions are known to be ( )
positive:

$$A(s) + B^{2+}(aq) \rightarrow A^{2+}(aq) + B(s)$$
$$A(s) + C^{2+}(aq) \rightarrow A^{2+}(aq) + C(s)$$

At standard concentrations, the reaction between B^{2+} and C:
    a) is spontaneous   b) is nonspontaneous
    c) is at equilibrium   d) cannot say

14. Given the following standard reduction potentials: ( )

$$Mn^{2+}(aq) + 2\,e^- \rightarrow Mn(s) \qquad\qquad -1.18 \text{ Volts}$$
$$2\,H_2O + 2\,e^- \rightarrow H_2\,(g) + 2\,OH^-(aq) \qquad -0.83$$
$$I_2\,(s) + 2\,e^- \rightarrow 2\,I^-(aq) \qquad\qquad\qquad +0.53$$
$$O_2\,(g) + 4\,H^+(aq) + 4\,e^- \rightarrow 2\,H_2O \qquad +1.23$$

we would predict that the electrolysis of a water solution of MnI_2 would probably produce:

a) Mn, I_2 b) Mn, O_2 c) H_2, I_2 d) H_2, O_2

15. For a certain redox reaction, E° is positive. This means that: ( )
   a) ΔG° is positive, K is greater than 1
   b) ΔG° is positive, K is less than 1
   c) ΔG° is negative, K is greater than 1
   d) ΔG° is negative, K is less than 1

16. For the reaction: $4\,Al(s) + 3\,O_2\,(g) + 6\,H_2O \rightarrow 4\,Al(OH)_3\,(s)$ ( )
n in the equation: $\Delta G^\circ = -nFE^\circ$ is
   a) 1   b) 2   c) 3   d) 12

17. For the redox reaction: $A(s) + B^{2+}(aq) \rightleftharpoons A^{2+}(aq) + B(s)$, ( )
$K = 10$. When the concentrations of B^{2+} and A^{2+} are 0.5 M and 0.1 M, respectively:
   a) the forward reaction is spontaneous
   b) the system is at equilibrium
   c) the reverse reaction is spontaneous
   d) cannot say

18. It is possible to increase the voltage of a cell in which the ( )
reaction is

$$Zn(s) + Cu^{2+}(aq) \rightarrow Zn^{2+}(aq) + Cu(s)$$

by
   a) increasing the concentration of Zn^{2+}
   b) increasing the concentration of Cu^{2+}
   c) increasing the size of the Zn electrode
   d) increasing the size of the Cu electrode

19. Which one of the following equilibrium constants would be ( )
most difficult to obtain from cell measurements?
   a) K_b for NH_3 b) K_c for $H_2\,(g) + Cl_2\,(g) \rightarrow 2\,HCl(g)$
   c) K_d for $Cu(NH_3)_4{}^{2+}$ d) K_w for water

20. A solution of NaOH saturated with Cl_2 at room temperature ( )
will contain appreciable concentrations of all but one of the following species. Indicate the exception.
   a) Cl^- b) OH^- c) ClO^- d) $ClO_4{}^-$

21. Which one of the following metals reacts with HNO_3 but not ( )
with dilute HCl?

    a) Pt   b) Mg   c) Na   d) Cu

22. Which one of the following metals reacts with dilute HCl but ( )
not with water?

    a) Ag   b) Na   c) Ni   d) Ca

23. In order to convert $CrO_4{}^{2-}$ to $Cr_2O_7{}^{2-}$, one would add ( )

    a) water   b) an acid

    c) an oxidizing agent   d) a reducing agent

24. Corrosion ordinarily occurs more readily in sea water than in ( )
fresh water because:

    a) Na^+ ions in the sea water attack iron

    b) Cl^- ions in the sea water attack iron

    c) sea water is a better electrical conductor

    d) O_2 is more soluble in sea water

25. "Aqua regia," a mixture of conc. HCl and conc. HNO_3, ( )
dissolves certain metals and metal sulfides which fail to dissolve in
conc. HNO_3. The main function of the HCl is to

    a) increase the conc. H^+

    b) furnish Cl_2, a better oxidizing agent than HNO_3

    c) furnish Cl^-, which acts as a complexing agent

    d) convert the metal to gold, which is soluble in HNO_3

SELF-TEST ANSWERS

  1. T
  2. F   (The reverse is true.)
  3. F   (Will depend on relative rates.)
  4. F   (Strong reducing agent.)
  5. F
  6. T
  7. T   (Write the Nernst equation.)
  8. T
  9. F
10. F
11. a
12. c
13. d
14. c
15. c
16. d

17. a
18. b
19. b   (Involves three gases.)
20. d
21. d   (Determine $E°$ for the possible reactions.)
22. c
23. b   (Can you write the equation?)
24. c
25. c

SELECTED READINGS

Taube, H., Mechanisms of Oxidation-Reduction Reactions, *J. Chem. Ed.* (July 1968), pp. 452-461.

A rather advanced discussion of the mechanisms of redox reactions involving transition metal complexes. (A supplement to the article by House, cited in Chapter 19).

For relating $E°$ to reaction spontaneity and free energy change, see the readings of Chapter 12.

22 NUCLEAR REACTIONS

QUESTIONS TO GUIDE YOUR STUDY

1. How are nuclear reactions different from "ordinary" chemical reactions? How would you experimentally recognize that a particular reaction was a nuclear reaction?

2. What factors determine whether a particular atom is radioactive? Are there correlations that can be made with the periodic table; with nuclear composition?

3. What are the properties of the various kinds of radiation? How, for example, do they interact with matter? (What chemical reactions occur as a result of interaction with biological systems?)

4. How do reaction conditions such as temperature, pressure and concentration affect the nature of a nuclear reaction?

5. How would you experimentally determine the rate of a nuclear reaction? What can you say about the rate law and reaction mechanism for a given nuclear reaction?

6. What can you say about the spontaneity and extent of a nuclear reaction?

7. What is the difference between "natural" and "artificial" radio-activity; between fission and fusion?

8. What are some of the *chemical* applications of nuclear reactions?

9. What energy effects are associated with nuclear reactions? Can you predict whether a given nuclear reaction will be exothermic or endothermic?

10. What reactions occur in a nuclear power plant? What are some of the advantages and disadvantages of nuclear power? (How, for example, does the cost compare with that of conventional power from fossil fuels?)

11.

12.

YOU WILL NEED TO KNOW

Concepts

1. How to symbolize (and describe) nuclear composition; i.e., how to use the notation that shows the atomic number and the mass number — Chapter 2
2. How to interpret or describe a reaction mechanism — Chapter 14
3. The general lay-out of the common long form of the periodic table — Chapter 6

Math

1. How to work problems involving the first order rate law — Chapter 14

CHAPTER SUMMARY—OBJECTIVES

In this chapter, we considered three different kinds of spontaneous nuclear reactions.

1) Radioactive decay, in which an unstable nucleus decomposes, emitting:

$$\text{an alpha particle: } {}^{238}_{92}\text{U} \rightarrow {}^{4}_{2}\text{He} + {}^{234}_{90}\text{Th}$$

$$\text{a beta particle: } {}^{234}_{90}\text{Th} \rightarrow {}^{0}_{-1}\text{e} + {}^{234}_{91}\text{Pa}$$

$$\text{or a positron: } {}^{30}_{15}\text{P} \rightarrow {}^{0}_{1}\text{e} + {}^{30}_{14}\text{Si}$$

and producing a new, more stable nucleus. Several steps may be required to form a nonradioactive nucleus. The decomposition of uranium-238 passes through 14 intermediates (8 α emissions, 6 β) yielding, as a final product, a stable isotope of lead, ${}^{206}_{82}\text{Pb}$.

A few radioactive isotopes, mostly those of the heavy elements, occur in nature. Many others have been produced in the laboratory by bombardment reactions using positively charged particles (protons, deuterons, α-particles), neutrons, or high energy radiation (X-rays, gamma radiation). An important accomplishment in this area has been the synthesis of at least one isotope each of elements of atomic number 93 to 105, the trans-uranium elements.

Radioactive decay follows the first order rate law (Chapter 14). Rates of different decay processes are often compared by citing half-lives, which can vary from a millisecond to many billions of years. Methods based on the decay of naturally occurring radioactive isotopes have been worked out to determine the age of rocks, both lunar and terrestrial, and carbon-containing artifacts.

2) *Fission,* in which a heavy nucleus, commonly $^{235}_{92}U$ or $^{239}_{94}Pu$, splits under neutron bombardment to give two lighter isotopes. A typical reaction is:

$$^{235}_{92}U + ^{1}_{0}n \rightarrow ^{90}_{37}Rb + ^{144}_{55}Cs + 2^{1}_{0}n$$

Fission ordinarily produces an excess of neutrons; provided a certain critical mass of fissionable material is present, a chain reaction can result.

3) *Fusion,* in which two light nuclei combine, e.g.,

$$^{2}_{1}H + ^{2}_{1}H \rightarrow ^{4}_{2}He$$

Processes of this type, unlike other types of nuclear reactions, have large activation energies. They occur at reasonable rates only at very high temperatures. As of this writing, the only feasible way to achieve and maintain these temperatures is by means of a fission reaction.

All nuclear reactions evolve large amounts of energy, most of it in the form of heat. The quantity of energy given off per unit mass of reactant increases in the order: radioactive decay $<<$ fission $<$ fusion. The energy change can be accounted for quantitatively by the Einstein relation:

$$\Delta E \text{ (in ergs)} = 9.00 \times 10^{20} \times \Delta m; \Delta E \text{ (in kcal)} = 2.15 \times 10^{10} \times \Delta m$$

Here, unlike ordinary chemical reactions, there is a detectable difference in mass between products and reactants. The enormous amounts of energy evolved in fission and fusion reflect the fact that the binding energy per nucleon is a maximum for isotopes of intermediate mass (Figure 22.6, p. 613). Since the plot of binding energy per nucleon vs. mass number rises very steeply near the origin and falls off more gradually at high mass numbers, considerably more energy is given off in fusion than in fission.

The reactions discussed in this chapter, none of which were known or even suspected a century ago, have had a profound effect upon our lives and our environment. A generation, born since the holocausts of Hiroshima and Nagasaki, has lived with the constant threat of a nuclear disaster that could destroy life on earth. Now, facing an energy crisis, we are becoming aware of the promise of nuclear reactions, particularly those of the fusion type, to supplement and eventually replace our dwindling supply of fossil fuels.

Objectives

After completing this chapter, you should be able to
1. write equations, balanced as to mass number and atomic number, to represent various nuclear reactions;

2. use the first order rate law to determine rates of decay and half-lives of radioactive isotopes, including those used to determine the ages of rocks and organic materials;

3. discuss the effects of radiation on human beings and materials;

4. explain the principles behind the use of radioactive isotopes in medicine, in industry, and in chemical research;

5. predict the mode of decay of certain unstable isotopes (Problem 22.12);

6. write typical equations to illustrate the fission and fusion processes and discuss the characteristics of these reactions;

7. use the Einstein relation to calculate ΔE for nuclear reactions, mass decrements and binding energies.

SELF-TEST

True or False

1. The emission of a β-particle leaves the atomic number   (   )
unchanged but increases the mass number by one unit.

2. The extent of deflection in an electrostatic field is greater for   (   )
a β-particle than for an α-particle.

3. Emission of a positron is equivalent to the conversion of a   (   )
proton to a neutron in the nucleus.

4. The most serious effect of low level radiation on the body is   (   )
that it produces severe skin burns.

5. The longer the half life of a radioactive isotope, the more   (   )
rapidly it decays.

6. The very heavy transuranium elements, atomic number 101   (   )
or greater, are most readily prepared by neutron bombardment.

7. The technique of activation analysis is limited to those   (   )
elements that have naturally radioactive isotopes.

8. According to the Einstein relation (Equation 22.19), the   (   )
fusion of one gram of deuterium would liberate 9.00×10^{20} ergs or
2.15×10^{10} kcal of energy.

9. Probably the most important hazard associated with a   (   )
nuclear power plant is the possibility of a nuclear explosion.

10. A plausible way to achieve the high temperatures required for   ( )
nuclear fusion is to operate the process in the upper atmosphere.
(Recall the temperature profile of the atmosphere described in
Chapter 15.)

Multiple Choice

11. The emission of an α-particle lowers the atomic number by   ( )
____ and the mass number by ____ respectively.
     a) 1, 1   b) 1, 2   c) 2, 2   d) 2, 4

12. Nuclear reactions differ from ordinary chemical reactions in   ( )
all but one of the following ways. Indicate the exception.
     a) The energy evolved per gram is much greater for nuclear
         reactions.
     b) Nuclear reactions occur much more rapidly.
     c) New elements are often formed in nuclear reactions.
     d) In nuclear reactions, reactivity is essentially independent
         of the state of chemical combination.

13. Emission of which one of the following leaves both atomic   ( )
number and mass number unchanged?
     a) positron   b) neutron   c) α-particle   d) γ-radiation

14. A certain radioactive series starts with $^{235}_{92}U$ and ends with   ( )
$^{207}_{82}Pb$. In the overall process, ____ α-particles and ____ β-particles
are emitted.
     a) 8, 6   b) 14, 10   c) 7, 10   d) 7, 4

15. Which one of the following instruments would be least suit-   ( )
able for detecting particles given off in radioactive decay?
     a) Geiger counter   b) scintillation counter
     c) electron microscope   d) cloud chamber

16. In determining the age of organic material, one measures   ( )
     a) the time required for half of the C-14 in the sample to
         decay
     b) the ratio of C-14 to C-12 in the sample
     c) the percentage of carbon in the sample
     d) the time required for half the sample to decay

17. The half life of uranium-238, which decomposes to lead-206, ( )
is about 4.5 × 10⁹ years. A rock which contains equal numbers of
grams of these two isotopes would be _____ years old.
a) less than 4.5×10^9 b) 4.5×10^9
c) more than 4.5×10^9 d) cannot say

18. Bombardment of $^{75}_{33}$As by a deuteron, $^2_1$H, forms a proton and ( )
an isotope which has a mass number of ____ and an atomic number
of ____ .
a) 73, 32   b) 75, 33   c) 75, 32   d) 76, 33

19. The element silicon has an atomic weight of about 28. The ( )
isotope $^{30}_{14}$Si would most likely decay by emitting a(n)
a) positron b) proton c) electron d) alpha particle

20. Which one of the following isotopes would be most likely to ( )
undergo fission?
a) $^{14}_6$C   b) $^{59}_{27}$Co   c) $^{239}_{94}$Pu   d) $^2_1$H

21. Which of the isotopes listed in Question 20 would you expect ( )
to have the largest binding energy per nucleon?

22. Which of the isotopes listed in Question 20 would be most ( )
likely to undergo fusion?

23. According to the Einstein relation (Equation 22.19), the ( )
energy given off in a nuclear reaction in which the decrease in mass is
2.0 mg would be
a) 1.8×10^{21} ergs   b) 1.5×10^{20} ergs
c) 9.0×10^{20} ergs   d) some other number

24. The masses of $^4_2$He, $^6_3$Li and $^{10}_5$B are 4.0015, 6.0135 and ( )
10.0102 amu, respectively. The splitting of a boron-10 nucleus to
helium-4 and lithium-6 would
a) evolve energy   b) absorb energy
c) result in no energy change   d) cannot say

SELF-TEST ANSWERS

1. F (Reverse is true.)
2. T (Much smaller mass.)
3. T
4. F
5. F
6. F (Heavier particles used.)

7. F
8. F   (Δm must be one gram.)
9. F
10. F   (Too few particles available.)
11. d
12. b   (Some are very slow — consider the range of half-lives.)
13. d
14. d   (Total mass change is due to seven α-particles.)
15. c
16. b
17. c   (Contains more moles of Pb.)
18. d
19. c
20. c
21. b
22. d
23. d   (1.8×10^{18}. How many kcal is this?)
24. b   (Δm is positive.)

SELECTED READINGS

Harvey, B. G., *Nuclear Chemistry*, Englewood Cliffs, N. J., Prentice-Hall, 1965.
A more general and detailed introduction to nuclear chemistry, about on the same level as the text.
Hershey, J., *Hiroshima*, New York, Knopf, 1946.
For fear that we may forget the potential misapplications of nuclear reactions . . . .
Hogerton, J. F., The Arrival of Nuclear Power, *Scientific American* (February 1968), pp. 21-31.
A survey of the development of nuclear power, to 1968, in its socio-economic context; with some forecasts.
Wahl, W. H. and Kramer, H. H., Neutron Activation Analysis, *Scientific American* (April 1967), pp. 68-82.
An important application of nuclear reactions, particularly useful recently in lunar analyses.

Also see the article by Seaborg cited in Chapter 6, and the books by Holdren and by Turk cited in Chapter 4.

23 AN INTRODUCTION TO BIOCHEMISTRY

QUESTIONS TO GUIDE YOUR STUDY

1. How would you isolate a pure sample of an enzyme or a nucleic acid? How would you test for purity? How would you determine the composition? (Are these procedures basically the same as those already encountered?)

2. How do biological molecules, like those of carbohydrates, proteins and fats, compare in molecular dimensions (e.g., volume and molecular weight) to each other and to the species dealt with so far?

3. What experimental methods are there for determining molecular structure (both the sequence of bonded atoms and their three-dimensional arrangement)?

4. What energy is available to an organism for driving its energy-requiring processes? How is it transferred? How is it stored?

5. What range of conditions (T, P, concentration . . .) prevail during ordinary biochemical reactions? What is known about reaction mechanisms?

6. Do proteins (and other complex biological molecules) spontaneously arise from amino acids (and other simpler molecular units)? Is life itself a spontaneous process?

7. Can we begin to explain the properties of living organisms in molecular terms? (Like growth and reproduction; mutations, disease, death.)

8. What kinds of answers can we now give to how life may have begun; how evolution occurs? (What kinds of questions cannot be, or at least have not been, answered?)

9. Why are there relatively *few* kinds of molecules important to life?

10. Can you now begin to account for the large-scale resolution of matter into simple gaseous substances (the atmosphere) solids and liquids (ores in the earth's crust, the oceans) and biological substances (as confined to living organisms) – a resolution that has occurred through eons and continues to occur?

11.

12.

YOU WILL NEED TO KNOW

Concepts

The application of chemistry to biological problems will prove more rewarding the more chemistry you know. Consider some of the concepts tied together in this chapter:
- methods of separation, e.g., chromatography – Chapter 1
- molecular weights – Chapter 2
- structural formulas, Lewis formulas – Chapter 7
- hydrogen bonding and the nature of other inter- and intramolecular forces – Chapter 7
- functional groups and their properties ($-NH_2$, $-CO_2H$, amide . . .) – Chapter 8
- polymers and their properties – Chapter 8
- methods and results of structure determination, e.g., x-ray diffraction – Chapter 9
- principles of solubility – Chapter 10
- catalysis – Chapter 14
- acid-base properties, pH – Chapters 17, 18
- oxidation state, oxidation-reduction – Chapters 20, 21

Math

Except for calculations in the problems involving MW, there is no math used in this chapter.

CHAPTER SUMMARY—OBJECTIVES

For all the sweat or tears of joy you may have shed by now, your own composition is still something like 65 per cent water, 15 per cent protein, 15 per cent fat, and less than 1 per cent carbohydrate – the rest being inorganic compounds. Despite the complexities of the thirty-odd per cent organic compounds and the reactions they take part in, we can note here several simplifications.

1. Though debate continues in the background, the nature of life seems to be explicable in terms of known principles of chemistry and physics. (In particular, no new chemical principles are needed in this chapter.) The molecular interpretation of biological structures and processes is already coming of age. For example: reaction pathways, the energy and material flows, have been discovered for most of the reactions involving the major constituents of living organisms.

2. Certain basic features are known to be common to all organisms: molecules with very similar architectures perform very similar functions, whether in one organism or another, plant, animal, or microbe. For example: the hemoglobin of man is built much like that of the horse and that of the gorilla, and all three serve to carry oxygen. There are really only a few kinds of biological molecules, and they are common to all life as we know it. (The usual classification lists carbohydrates, fats, proteins and nucleic acids as major components.)

3. Much of the chemistry of life occurs in dilute aqueous solution (and so rules out the need for studying many kinds of substances incompatible with water), at or near room temperature. The reactions generally, if not always, involve enzyme catalysis, and often oxidation-reduction occurs.

4. Many, if not most, reactions are interlocked (coupled) with others. The energy released in one reaction is, at least in part, used to drive another reaction. For all coupled reactions, there is a net decrease in free energy. There is no creation of order in the living organism that doesn't occur simultaneously with a larger disordering of the surroundings.

In addition to being one of the most active research areas in recent years, molecular biology has been perhaps *the* area of cooperation among the various "compartments" of science. Working for perhaps a dozen man-years in elucidating the amino acid sequence for a single small protein, a team of physicists, chemists and biologists have freely shared the tools and theories perfected individually, while training one another in the problems of their new frontier.

What are some of the unresolved or incompletely solved problems?

1. *Molecular structures* are known in three-dimensional detail for only a handful of biological species. We know most other structures only in sketchy outline. All evidence so far indicates that all the biological properties of a molecule are determined by its primary structure (the number and kinds of atoms present and their bonding sequence).

2. *Molecular functions* and reaction mechanisms have been vaguely described for just a few representative compounds. Again, the details need to be discovered before a thorough understanding can be achieved.

3. *Comparative studies* of molecules serving similar functions need to be pursued in order to further describe the mechanism by which evolution

occurs, by which only a few types of molecules seem to have survival value. The current work on compounds associated with fossils, as well as the abiological syntheses of compounds associated with living organisms, should continue to shed light on the possible origins of life itself.

4. Humanistic and moral questions need be raised: most biochemists consider as inevitable deliberate *design* of organisms, as well as modification of organisms already alive. All of us need to know where we may be going.

Objectives

After working through this chapter, you should be able to:

1. describe the typical composition and approximate dimensions (mass, molecular weight, volume) of carbohydrate, protein, and nucleic acid molecules;

2. describe the structure of these biological molecules, distinguishing between monomers and polymers and between primary, secondary and higher levels of organization (monomer sequence, helix formation, conformation . . .);

3. account for some of the characteristics of these structures (and changes in structure resulting from changes in conditions such as pH, T, . . .) in terms of inter- and intramolecular forces;

4. describe some of the functions of these molecules (e.g., the highly specific nature of enzyme action);

5. recognize the potential for optical activity in small molecules;

6. trace, in general terms of structures and reactions, energy and material transfers from environment to organisms, within organisms, and from organisms to the surroundings (see, for example, number 4 in Questions to Guide Your Study);

7. describe in general terms the storage and use of genetic information (code);

8. discuss some chemical evidence which has been taken as supporting evolution.

SELF-TEST

True or False

1. In animals, proteins serve primarily to store energy. ( )

2. Carbohydrate polymers are synthesized only by plants. ( )

3. The net photosynthetic reaction,   ( )
$CO_{2(g)} + H_2O_{(g)} \rightarrow glucose_{(s)} + O_{2(g)}$, involves a reduction of carbon.

4. The α-amino acids might generally be expected to form   ( )
chelates with metal atoms.

5. The water solution of any amino acid which contains only   ( )
one basic functional group and one acidic group is expected to have a
pH of 7.

6. The hydrolysis of proteins (breakage of the peptide links   ( )
with insertion of water molecules) will always result in the formation
of equal numbers of $-NH_2$ and $-CO_2H$ groups.

7. The formation of an enzyme from amino acids has a positive ( )
free energy change. The fact that the synthesis of enzymes occurs in
a living organism is a contradiction of the laws of thermodynamics.

8. Cellular RNA can be thought of as an exact copy of DNA.   ( )

Multiple Choice

9. Where rigidity in structure is required (as in an animal's ( )
skeleton), the material is likely to consist of
    (a) DNA (b) nucleotides (c) polysaccharrides (d) alkanes

10. Which one of the following formulas would you choose to   ( )
represent an average composition of protein?
    (a) CH_2O (b) $C_{57}H_{110}O_6$ (c) CH_7NO (d) $C_9H_{13}SN_2O_2$

11. The largest amount of heat would be obtained from the   ( )
complete combustion of a gram of:
    (a) protein   (b) carbohydrate
    (c) fat (e.g., $C_{57}H_{110}O_6$)   (d) DNA

12. The geometry associated with the four atoms of the peptide ( )
    O H
    ‖ |
link, $-C-N-$, is observed to be:

    (a) linear   (b) tetrahedral   (c) planar
    (d) variable, depending on the rest of the chain

13. The method of choice for determining the molecular weight ( )
of a protein is:
    (a) osmotic pressure
    (b) gas density
    (c) freezing point lowering
    (d) direct weighing of a single molecule

14. The best solvent for a polypeptide is likely to be   ( )
    (a) CCl_4 (b) CH_3-O-CH_3 (c) H_2O (d) HF

15. To separate a mixture of monosaccharides, you would   ( )
probably use a(n):
    (a) centrifuge   (b) column chromatograph
    (c) mass spectrometer   (d) electrolytic cell

16. In a 0.10 M solution of glycine, $H_2C(NH_2)CO_2H$, at a pH of ( )
10, the most abundant species next to water is
    (a) OH^- (b) $H_2C(NH_3)CO_2$
    (c) $H_2C(NH_3)CO_2H^+$ (d) $H_2C(NH_2)CO_2^-$

17. The linkage between monomers ("residues") in a nucleic acid ( )
involves
    (a) C-N   (b) C-O-C   (c) P-O-C   (d) hydrogen bond

18. The maximum number of different tripeptides that can be ( )
formed from three different amino acids, using one residue of each,
is:
    (a) 2   (b) 3   (c) 4   (d) 6

19. The maximum number of different tripeptides that can be ( )
made from three different amino acids, using any number of residues
of each, is:
    (a) 2   (b) 3   (c) 6   (d) 27

20. A synthetic RNA made up of only the base uracil would ( )
probably cause a cell to produce a polypeptide which contained only
    (a) uracil
    (b) phenylalanine
    (c) three phenylalanines
    (d) any amino acid other than Glu, Gln or Lys

21. As the temperature is increased, the rate of an enzyme- ( )
catalyzed reaction first increases and then decreases. The decrease
can be explained in the following way:
    (a) all reactions achieve a maximum rate at some tempera-
        ture
    (b) the molecules acted on by the catalyst spontaneously
        decompose at the high temperature
    (c) intramolecular forces (e.g., H-bonding) in the enzyme
        molecule begin to break down
    (d) all catalyzed reactions behave in this unaccountable
        manner

22. The rate of an enzyme-catalyzed reaction is generally found ( )
to increase with the concentration of the reactant substrate up to
some maximum rate and then level off. This levelling-off is probably
due to:

    (a) decomposition of enzyme
    (b) a change in enzyme conformation
    (c) molecules of reactant get in the way of each other at high
        concentrations
    (d) all the "active sites" on the enzyme molecules are
        occupied

23. Which one of the following species would you expect *not* to ( )
show optical activity?

    (a) $CH_2(NH_2)CO_2H$ (b) $CH_3CH(NH_2)CO_2H$

    (c) $CH_3\overset{\overset{OH}{|}}{C}H-\overset{\overset{O}{\|}}{C}H$ (d) $CH_3CH_2\overset{\overset{OH}{|}}{C}HCH_3$

24. A change in the composition of which kind of molecule is ( )
most likely to result in a mutation?

    (a) RNA   (b) DNA   (c) carbohydrate   (d) enzyme

25. The "backbone" in a protein molecule can assume only a ( )
limited number of conformations. This is a result of:

    (a) hydrogen bonding
    (b) relative solubilities of attached groups
    (c) the geometry of the peptide link
    (d) all of the above

26. The argument for evolution gains support from which of the ( )
following?

    (a) proteins, nucleic acids and carbohydrates are not
        sufficiently stable to be found in fossils
    (b) the function of a particular enzyme in man is served by
        enzymes of similar structure in other organisms
    (c) the "active site" in an enzyme is the locale for most
        mutations
    (d) DNA is found in all organisms

SELF-TEST ANSWERS

    1. F   (Fats and glycogen serve in this capacity; proteins, in many
          others.)
    2. F
    3. T

4. T (Like:
$$\begin{matrix} \overset{|}{C}\text{–O} \\ | \qquad M) \\ \text{–}\overset{|}{C}\text{–N} \end{matrix}$$

5. F

6. T

7. F (First, the biosynthesis doesn't begin with free amino acids but with complex molecules such as proteins. Second, other reactions occur that supply the free energy necessary for the synthesis.)

8. F

9. c (Where the most H-bonding can occur. However, carbohydrates serving this function are mainly found in plants; proteins, in animals.)

10. d

11. c (Carbon in lowest oxidation state. Contrast C, CO and CO_2 as fuels.)

12. c

13. a (Most sensitive.)

14. c (A nondestructive solvent for these molecules that contain lots of polar groups.)

15. b

16. d (What is $[OH^-]$?)

17. c

18. d (Each amino acid residue could be at the $-NH_2$ end or at the $-CO_2H$ end.)

19. d

20. b

21. c

22. d

23. a (No asymmetric carbon.)

24. b

25. d

26. b

SELECTED READINGS

Calvin, M., *Chemical Evolution*, New York, Oxford University Press, 1969.
   A rather technical and personal discussion that offers much to the reader, whether skimmed or read carefully. Discusses chemical fossils as well as work done in laboratory attempts at synthesizing biological molecules and systems.

Handler, P., Ed., *Biology and the Future of Man*, New York, Oxford University Press, 1970.
   Intended for the well-read "layman"; there is much chemistry here as well as an overview of research in biology in general.

Watson, J. D., *Molecular Biology of the Gene* (2nd edition), Menlo Park, Calif., W. A. Benjamin, 1970.
A very well-written account of much of modern biochemistry, by one who has pioneered.
Chemical and Engineering News has carried several interesting feature articles relevant to this chapter. Some examples:
Chemical Origins of Cells (June 22, 1970 & Dec. 6, 1971)
Some Chemical Glimpses of Evolution (Dec. 11, 1967)
Life Transcending Physics and Chemistry (Aug. 21, 1967)
The Synthesis of Living Systems (Aug. 7, 1967)
Scientific American is a very rich resource. Consider about half the biochemical articles published in 1972 alone:
The Carbon Chemistry of the Moon (October)
The Chemical Elements of Life (July)
Organic Matter in Meteorites (June)
The Structure and History of an Ancient Protein (April)
RNA-Directed DNA Synthesis (January)